The Axe Estuary Bird Report 2007

The Axe Bird Group

Copyright © 2008 East Devon District Council, Countryside Service
First published May 2008

Published in 2008 by East Devon District Council, Countryside Service
Knowle, Sidmouth, Devon, EX10 8HL
Tel: +44 (0)1395 517557
E-mail: countryside@eastdevon.gov.uk
Web: www.eastdevon.gov.uk/countryside

Acknowledgements
The editorial committee would like to extend their heartfelt thanks to the following, without whose help this report could not have been produced:

All contributors to the LNR logbooks and BirdForum's 'Backwater Birding' blog, and to local birders Phil Abbott, Kevin Hale, David Walters and Karen Woolley, for access to their personal records.
Photography: Roger Boswell, Andrew Coates, Gavin Haig, Richard Halliwell, Bob Hastie, Mike Hughes, Mike Lock, Stuart Piner, Fraser Rush, Steve Waite, Karen Woolley.
Illustrations: Mike Hughes
We would also like to thank East Devon District Council's Xerox Print Room for their expertise in designing this report.

Editor: Gavin Haig

Contributors:
Donald Campbell   Fraser Rush
James Chubb   Michael Tyler
Ian McLean   Steve Waite
Bob Olliver

Printed by: Inprint Litho Limited, Unit 4, Stanton Road Industrial Estate, Stanton Road, Southampton, SO15 4HU

Printed on recycled paper using environmentally friendly non soya based ink

All rights reserved. No part of this publication may be reproduced in any way without the written permission of the publishers.

ISBN 978-0-9558706-0-6

The Axe Estuary and Seaton Bay Bird Report 2007

CONTENTS

2007 Editorial	2
Review of the Year	3 - 11
Species Accounts	12 - 76
Species List	77 - 80
Maps of Report Area	81 - 85
The MSC Napoli Incident	86 - 88
Axe Estuary Ringing Group Report 2007	89 - 90
Dragonflies and Damselflies (*Odonata*)	91 - 94
Butterflies	95 - 96
Audouin's Gull at Seaton Marshes	97 - 99
Bonaparte's Gull on the River Axe	99 - 100
Iberian Chiffchaff at Beer Head	100 - 102
Stone-curlew at Seaton Marshes	102 - 103
East Devon District Council Local Nature Reserves	104 - 109
Timed Tetrads and Roving Records	110 - 112
Contacts and Links	113
Websites of Interest	114

The Axe Estuary and Seaton Bay Bird Report 2007

2007 Editorial – Gavin Haig

Early 2007 saw the publication of the first Axe Estuary and Seaton Bay (AESB) Bird Report – a result of the eager and willing cooperation of local birders, the hard work of the editorial committee and the backing of East Devon District Council. Although well received there was room for improvement, and we hope this second edition reflects our attempts to address that opportunity. We have been grateful for constructive criticism and helpful suggestions and would like to show that we were listening! For example, photographs have now been spread through the report and will generally appear next to the relevant species account; there are a number of finder's accounts – short articles written by (usually) the finder of a rare bird – as well as articles on other topics of local interest, including dragonflies, damselflies and butterflies. Finally, we have gratefully bowed to the suggestion that we desist from using British Ornithologist's Union (BOU) recommended species names (some of which may be unfamiliar or confusing to many British birders) and revert to those in common use.

A major event of the year was the grounding in January of the damaged container ship MSC Napoli, just off Branscombe. The effect on local (and not-so-local) birds was dramatic, and is examined in an article in this edition.

Bird-wise, 2007 will stand out as the best year ever for rare and scarce species, with the quality of birding to be enjoyed in this tiny recording area exceeding most of our expectations. Many whole counties would envy us our species list this year! Birding is not all rarities and scarcities though, and we have endeavoured to pay a little more attention to documenting our breeders too.

It is worth mentioning that this year's report was compiled from data which included around 4,000 records culled from the notebooks of local observers! It is probably fair to say that 'the best year ever' may relate as much to observer effort as to the actual occurrence of obliging birds – they are presumably always available, if we only look for them.

Finally, we must mention the Devon Bird Report published by Devon Bird Watching and Preservation Society– an invaluable source of reference. We neglected to do so last year and would like to put that right. Without the existence of that august publication this report would never have been conceived. As it is, we have shamelessly plagiarised many aspects of its content and layout and are extremely grateful for its influence!

The Axe Estuary and Seaton Bay Bird Report 2007

Review of the Year – Steve Waite

* indicates bird records that are pending consideration by either the Devon Birds Records Committee or British Birds Rarities Committee at time of going to press.

The recording area probably received even better coverage than in 2006, resulting in a superb total of **211** species being recorded. There were some excellent highlights, including the addition of eight new species to the area list, which now stands at **269** (pending the acceptance of all rarity records). These included four national rarities: **Black Stork*** (2), **Bonaparte's Gull**, **Audouin's Gull** and **Iberian Chiffchaff**; the others were **Honey Buzzard** (3), **Temminck's Stint**, **Caspian Gull** and **Lapland Bunting**. Amongst many other highlights it is worth noting that **Little Egret** was confirmed as breeding for the first time. Picking a single 'event of the year' is always going to be subjective, but the occurrence of another national rarity – **Laughing Gull** – plus **Ring-billed Gull** (2), **Iceland Gull** (6) and **Glaucous Gull** (3) contributed to a gull list comprising an astonishing 16 species! 2007 will probably be remembered as 'The Year of the Gull'!

January

The mild winter continued, which meant that many common bird numbers were low, but this month was still very eventful, aided by the fact that last year's December star bird - the **Surf Scoter** - remained off Beer until the beginning of March, along with the site-faithful **Black Redstart**. A first-winter **Little Gull** on the estuary on 1st was a surprise, along with three **Mediterranean Gulls**; on 2nd a drake **Pochard** graced Lower Bruckland Ponds with an adult **Yellow-legged Gull*** on the estuary on 3rd. A **Black-necked Grebe** briefly on 4th off Beer was later seen flying past Branscombe, along with a **Black-throated Diver**, and on the same day a male **Hen Harrier** flew back and forth over Colyford Marsh. The quality continued on 6th, with a first-winter **Iceland Gull** on the river for the afternoon, leaving to roost on the sea at dusk. A lone **Brent Goose** sat on the sea on 9th and a couple of wintering **Green Sandpipers** became regulars at Colyford Common. A female **Pintail** and drake **Gadwall** were on Colyford Marsh on 10th, and the same day showed a good collection of divers off Branscombe with 17 **Red-throated** and one **Black-throated**. A **Firecrest** showed up at Jubilee Gardens in Beer on 13th, and two **Brent Geese** were on Colyford Marsh from 16th, remaining till 21st. Seawatching on 18th revealed two **Grey Plovers** west, along with good numbers of **Gannets**, **Fulmars** and **Kittiwakes** and, in the evening, the first-winter **Iceland Gull**, or another one, showed up briefly on the estuary. It remained on and off for the remainder of the month. The tragic events concerning the MSC Napoli made birding rather unpleasant for a short while, but a concerted effort was made as cold weather had finally set in. Three **Brent Geese** were on Seaton Marshes on 22nd and 49 **Dunlin** – a high count by recent standards – were on the river on 25th. The following day produced a couple of quality waders: a male **Ruff** on Colyford Marsh and a **Knot**, which remained for the rest of the month on the estuary. Two **Slavonian Grebes** fed briefly off Seaton on 27th, and seven **Brent Geese** flew west. On 29th, seven **Woodcock** were near Colyton and a **Dipper** was seen on the Coly near Umborne Bridge.

February

The cold spell continued and produced a single **Golden Plover** on 2nd. 12 **Brent Geese** flew east past Seaton on 4th and, on 5th, a **Greylag Goose** flew in off the sea and remained in the area for several weeks. Mild weather then arrived which ensured a quiet spell until what was probably a new first-winter **Iceland Gull** was found on 15th, with six **Water Pipits** at Colyford Marsh the next day – the highest count of the first winter period. A drake **Goosander** on the estuary on 17th was a nice surprise, as was a flock of 60 **Brent Geese** over the sea the following day. The 19th proved a busy day for gulls, with the discovery of a second-winter **Ring-billed Gull**, along with the **Iceland Gull** and at least eight **Mediterranean Gulls**. A couple of unseasonable **White Wagtails** were on Colyford Marsh from 21st; also here, several *littoralis* **Rock Pipits** were developing into summer plumage. Except for a **Velvet Scoter** flying west on 24th, the main focus for the final week of the month was gulls, with some heavy passage taking place at sea and on the estuary – this included at least twenty **Mediterranean Gulls** and thousands of **Common Gulls**. The top prize waited till 28th though, with a first-winter **Laughing Gull** off the seafront briefly, with an adult **Little Gull**. A **Dipper** on Colyford Common on this date was an unusual record.

March

The month started well with a male **Dartford Warbler** singing on Colyford Common, but the quality gull theme continued on 4th, with a first-winter **Glaucous Gull** and two adult **Little Gulls** on the river. The following day, last month's second-winter **Ring-billed Gull** reappeared on the river, along with a new **Iceland Gull** and, on 6th, a male **Pochard** fed here. There was another rare gull on 7th when an adult **Ring-billed Gull** was found mid-afternoon and a **Firecrest** appeared at Jubilee Gardens. With high pressure taking over, the first signs of spring brightened things on 8th, with a male **Wheatear** on Beer Head late in the day, followed by a **Sand Martin** over Seaton Marshes on 10th; an **Iceland Gull** on the estuary on this date appeared to be yet another new individual! Five **Pale-bellied Brent Geese** (along with one **Dark-bellied**) were sat on the sea off Seaton on 11th slowly drifting west and two **Black-throated Divers** were settled off Branscombe on 13th. With the blue skies and sunshine continuing an unusual raptor was on the cards, and on 16th it showed up in the form of a **Red Kite** over Beer. Cold northerlies then set in, but this didn't stop the first **Swallow** showing up on 19th. Things did go quiet for a while until a **Little Ringed Plover** dropped in at Seaton Marshes on 27th. Two **Iceland Gulls** were on the estuary on 28th, with a first-winter **Glaucous Gull** the following day, and a **Yellow-legged Gull*** on 30th and 31st. Two **Firecrests** were in Beer on 30th and, before the month's end, the year's first **Willow Warbler**, **House Martin** and **Sandwich Tern** were noted.

April

The first **Manx Shearwater** passed on 1st, and on the afternoon of 2nd Seaton Marshes was the place to be with singles of **Osprey** and **Merlin**, with a **Little Ringed Plover** on the marsh. A male **Tufted Duck** was at Lower Bruckland Ponds on 5th and a **Red Kite** flew over Seaton the following day. The year's first **Redstart** showed up on Beer Head on 7th and the 8th produced a second **Egyptian Goose** at Colyford Marsh (only a single bird had been present since the first week of January), with a **Whimbrel** and **Greenshank** here on 10th and the first **Lesser Whitethroat** on 11th. The 13th was a magic day, primarily due to the appearance of the patch's second ever **Stone-curlew** at Seaton Marshes, which remained all day, but also with the supporting cast of two **Little Ringed Plovers**, a **Little Gull** and a **Goosander** around the river, a male **Tufted Duck** at Lower Bruckland and a **Ring Ouzel**, a **Pied Flycatcher** and two **Redstarts** at Beer Head. Waders became more of a feature from mid-month, with **Bar-tailed Godwits**, **Dunlin** and **Ringed Plover** all noted on many dates, also the more usual summer migrants, with several **Grasshopper Warblers**, **Redstarts** and ones and twos of the other usual species recorded before the month's end. A pair of **Gadwall** flew west on 16th and on 17th two **Grey Partridges** crossing a main road were a bit of a shock for one observer – presumably the same pair was seen again the next morning at Haven Cliff, and on the same day a **Greylag Goose** was a brief visitor to Colyford Marsh. A **Cuckoo** on 19th at Beer Head didn't hang around, neither did a flock of 22 **Eider** which flew west past Branscombe on 21st. The first **Hobby** was seen on 23rd and another **Cuckoo** was heard at Branscombe the following day, though this was totally eclipsed by the appearance of three **Little Terns** off the seafront – the first confirmed record of this species on the patch for about 40 years! With a hint of onshore breeze, singles of **Great** and **Arctic Skua** passed Seaton on 25th and a summer plumaged **Red-throated Diver** sat off here the next day. A **Hoopoe** spent the day in private gardens on the outskirts of Beer on the 26th (eluding all the local birders!) and a drake **Garganey** on Seaton Marshes on 28th set things up nicely for an unbelievable finale to the month, with two firsts for the area! Firstly, also on 28th, a male **Iberian Chiffchaff** sang on Beer Head for the afternoon, and on 30th, a first-summer **Bonaparte's Gull** was found on the estuary, this also remaining for the afternoon.

May

A **Little Stint** on the river on the first day of the month was still present the following day when it was joined by a **Little Gull** and, more notably, a **Little Tern**. A drake **Garganey** at Seaton and Colyford Marsh on 3rd was a lot more appropriate to the season than four **Greylag** and two **Barnacle Geese** – this out-of-season goose flock remained till 7th. A spell of windy and showery weather came over during the middle of the month, which brought about some good movements on the sea, including several **Great Northern Divers**, two **Black-throated Divers**, seven **Storm Petrels**, five **Pomarine Skuas**, two **Great Skuas**, 15 **Arctic Skuas** and some quality tern sightings with a single **Roseate Tern** on 11th and two **Arctic Terns** the day before. The large numbers of waders on the move over the sea (mostly **Whimbrel**, **Dunlin** and **Sanderling**) were reflected on the river with an amazing 181 **Dunlin** present here on the 10th, and flocks of over 90 present over the next few days, with the odd **Sanderling**, **Ringed Plover** and a **Greenshank** on 13th. A pelagic trip from Beer which ventured no more than four miles out on 16th produced an amazing 116 **Storm Petrels**, a **Puffin**, three **Pomarine** and four **Arctic Skuas** and, rather oddly, four **Turnstones**! The wind picked up again, producing even more **Great Northern Divers**, **Storm Petrels**, **Pomarine Skuas** and several more waders including a **Greenshank**. Away from the sea, some quality wader sightings from the river included single **Curlew Sandpipers** on 22nd and 24th, and a major surprise on the afternoon of 23rd with a **Pectoral Sandpiper**. A female **Pochard** at Colyford Marsh on 30th was unseasonal.

June

Continuing the unprecedented spring for this species on patch, a single dark phased **Pomarine Skua** passed the seafront on 3rd. An amazing discovery on 5th, although ever so slightly off patch, was a **Black Stork*** in Southleigh. News was slow to reach the birding community, and it wasn't until 8th when two very privileged birders saw it just after dawn. Although it wasn't seen again by birders, it was still in the vicinity early on 9th when it was seen within the area covered by this bird report. A **Pale-bellied Brent Goose** on the river from 7th was somewhat out of season and remained for over a week. A few small waders were still popping up on the estuary with **Ringed Plover** and **Dunlin** present on a few dates, and a **Sanderling** on 12th. The first **Balearic Shearwater** flew east on 15th with a few more over the next few days along with good numbers of **Manx Shearwaters**, half a dozen **Storm Petrels** and a handful of **Arctic Skuas**. Away from the sea, a **Honey Buzzard** over Colyford Common early on 18th was a nice surprise, and the autumn's first **Green Sandpiper** was present, although numbers of these soon increased with eight here by 22nd with the season's first **Common Sandpiper** the following day. The autumn's first five **Black-tailed Godwits** were at Colyford Marsh on 25th and an unseasonal surprise on 27th was a female **Goldeneye** on the river. With **Storm Petrels** and good numbers of shearwaters still offshore, on the same day the season's second pelagic trip from Beer was arranged, with 40+ **Storm Petrels**, 168 **Manx Shearwaters**, two **Arctic Skuas** and a **Sanderling** the rewards. On the last day of the month a **Firecrest** was an unexpected find in woodland near Colyton, perhaps a local breeder?

July

With unsettled windy weather continuing, the first day of the month produced singles of **Little Gull** and **Eider** over the sea, along with the usual array of **skuas**, **shearwaters** and **Storm Petrels** – these featuring throughout the first week of the month. Once the weather died down, most eyes were focused on the estuary and LNRs with waders on the move. A juvenile **Little Ringed Plover** remained on Colyford Marsh for a few days from 9th, with the first returning **Dunlin** here on 14th – the same day as the first juvenile **Mediterranean Gull** of the season (a day after the first adult bird of the autumn was seen). A **Tufted Duck** flew downriver on 15th and a juvenile **Yellow-legged Gull*** was found on the estuary on 16th, with a **Turtle Dove** in Beer also on this date. A blow on 17th produced three **Balearic Shearwaters** and the first **Wheatear** of the autumn was on Beer Head the following day. A female **Garganey** showed up late in the day on 20th at Colyford Marsh, but didn't stay long, neither did the autumn's first **Osprey**, staying for the morning of 24th only. With a southerly wind picking up again for another few days, another pulse of **Arctic Skuas** and **Manx** and **Balearic Shearwaters** passed over the sea on 25th/26th. Another couple of juvenile **Yellow-legged Gulls*** were on the estuary over the last few days of the month, and a young **Dartford Warbler** at Haven Cliff on 30th was unexpected.

August

A brief **Wood Sandpiper** on Colyford Marsh started the month off on 1st, then the following day another **Yellow-legged Gull*** was more predictable than a **Honey Buzzard** which flew over Colyton on 3rd. A few more **Balearic Shearwaters** passed on 4th with yet another **Yellow-legged Gull*** on the river (with several more individuals throughout the month). Three **Sanderling** at Haven Cliff on 6th looked out of place over a crop field, though not as out of place as a **Black Stork*** which flew low north over Seaton on 8th – eluding all the local listers. Beer Head was busy on 11th with three **Dartford Warblers** (these birds probably responsible for various sightings over the follow weeks), two **Whinchats** and four **Sedge Warblers**. A **Grasshopper Warbler** at Colyford Common on 13th was a good autumn record, but no one could foresee what was just around the corner, with the astonishing discovery of an adult/sub-adult **Audouin's Gull*** on fields neighbouring Seaton Marshes on 14th. It disappeared early afternoon but was seen for one last time just prior to dusk in the town. The morning of 15th saw an unprecedented passage of **Great Skuas**, with a total of ten passing west in one hour, plus an **Arctic Tern**, also a **Sanderling** on the beach. The autumn's first **Redstart** graced Beer Head on 17th and with strong

southerly winds, seawatching on 18th produced an adult **Black** and two **Arctic Terns**, two **Great Skuas** and six **Balearic Shearwaters** with a **Knot** on the river. An **Osprey** flew south over Seaton on 19th, and 23rd was a good day for waders with single **Ruff**, **Wood Sandpiper** and **Turnstone** all seen, along with two early returning **Wigeon**. The 24th showed an increase in **Yellow Wagtails**, with 40 on Beer Head plus a **Redstart**, **Whinchat** and **Garden Warbler**, with other singles of **Whinchat** and **Redstart** behind Beer Cemetery. More migrants dropped in on Beer Head and behind Beer Cemetery the following day, including three **Pied** and nine **Spotted Flycatchers**. Another **Osprey** flew over Seaton on 25th and more migrants at Beer Head on 26th included another **Pied** and ten **Spotted Flycatchers** and a **Grasshopper Warbler**. On 28th two more **Ospreys** passed over the area (an early morning bird over the sea and one late afternoon south over town) and a second **Wood Sandpiper** dropped in at Colyford Marsh. Three **Curlew Sandpipers** appeared on 29th at Colyford Marsh, all remaining into September, and yet another **Osprey** passed over the area on 30th, though this one flew north!

September

A few more juvenile **Yellow-legged Gulls*** and a third **Wood Sandpiper** at Colyford Marshes were not unexpected during the start of the month, but a **Honey Buzzard** over Haven Cliff on 2nd certainly was! An **Osprey** fishing over the river on 4th made the Axe its home till the middle of the month; it was last seen when a second bird appeared. A **Turnstone** and a **Ruff** were a couple of nice highlights on the estuary on 5th, and a new **Curlew Sandpiper** appeared on 7th. A **Pied Flycatcher** on Beer Head on 9th was followed by at least three more later in the month behind Beer Cemetery. Another few **Curlew Sandpipers** and the autumn's first **Golden Plover** over Beer Head on 10th were yet more wader highlights. An adult **Arctic Tern** on Colyford Marsh on 11th was the first for the reserve, and a **Goosander** on the estuary later in the day was unseasonal. Two **Little Stints** on Colyford Marsh on 12th were eclipsed by the finding of a **Temminck's Stint** here the following day – it remained till the month's end. A **Pied Flycatcher** on Colyford Common on 15th was a quality reserve rarity, and a blow the following day encouraged two **Balearic** and one **Sooty Shearwater** to pass offshore. Another or the same **Goosander** landed on the river on 17th when the **Curlew Sandpiper** count had risen to six. More seawatching was required with strong winds blowing from the south. This produced a record number of **Balearic Shearwaters** – with 64 on 20th, along with two **Sooty Shearwaters**, several **Arctic Terns** and early **Red-throated** and **Great Northern Divers** and a **Brent Goose**. Also on this date a **Bar-tailed Godwit** and a **Turnstone** dropped in on the estuary. The 21st saw another **shearwater** record broken, with seven **Sooty Shearwaters** past during the morning, along with a few **Balearics**. A **Corncrake** was a great surprise at Colyford Common on 24th, and a **Grey Plover** seen here from 27th remained for at least a week. A male **Marsh Harrier** hunting over Colyford Marsh during the evening of 29th would have been even more of a highlight if the local birding network had found out sooner; still, a smart addition to the year list! After a few **Redstarts** on Beer Head during the month, a rather attractive male on 30th rounded the month off nicely along with a female **Ring Ouzel** and a **Merlin**.

October

The month started with an overdue year-tick in the form of a **Lesser Redpoll** over Colyton, but a late **Wood Sandpiper** which remained for a week from 2nd was more unexpected. A **Yellow-browed Warbler*** on Beer Head brightened up the day for one birder on 3rd, with a **Merlin** over Haven Cliff far more routine. A **Short-eared Owl** flew over Colyford Common on 5th, and a **Dartford Warbler** was on Beer Head. A **Spotted Redshank** on 6th was joined by a second bird the following day and remained on Colyford Marsh till 9th, with yet another **Curlew Sandpiper** appearing on 8th which, along with a **Jack Snipe** on Haven Cliff and the array of others on Colyford Marsh, produced a wader day list of 17 species! Another **Yellow-browed Warbler**, this time at Lower Bruckland Ponds from 9th was welcome as it remained till 11th, when a surprise **Whooper Swan** flew over. Three **Bearded Tits** at Colyford Marsh late on 12th were all too brief, though compensation came on the 14th with a **Cattle Egret** which roosted each night on Seaton Marshes till 21st when it flew west over Beer Head. A **Little Gull** on Colyford Marsh on 16th was a direct result of the poor weather, then from 17th a **Velvet Scoter** took up residence off Seaton Hole, being joined by a second bird from 25th. The autumn's first **Black Redstart** turned up in Beer, also on 16th, the first of a few individuals dotted around on patch over the next few weeks. A very late **Wood Sandpiper** showed at Colyford Marsh on 17th – yet another wader highlight for the reserve! A **Pallas's Warbler** on 22nd opposite Coronation Corner remained frustratingly elusive till it disappeared the following morning – though a **Firecrest** did ease the pain for some. At least two **Ring Ouzels** were on Beer Head on 25th, with three and a **Firecrest** the next day. Unsettled weather followed, which meant gull numbers increased drastically on the river, with the highlight being the area's first **Caspian Gull**, a second-winter which turned up alongside three **Yellow-legged Gulls***. A new **Yellow-legged Gull*** appeared the following day on Colyford Marsh, and seawatching produced 16 **Brent Geese**, a **Red-breasted Merganser** and two late **Balearic Shearwaters**. The weather soon cleared, and clear skies at the end of the month ensured overhead passage moved up several gears, with **Woodpigeon** counts of 11,150 on 29th and 22,881 on 30th. Other birds mixed up in the movement over the last few days of the month included singles of **Hen Harrier** and **Lapland Bunting**, two **Merlin**, several **Golden Plovers** and large numbers of finches including several **Brambling**, **Redpoll**, **Siskin** and a huge count of 2,307 **Chaffinches**. Grounded highlights included a **Tufted Duck** on Borrow Pit from 29th, with two more on the sea on

31st and another wing-barred *Phylloscopus* warbler – the month's third **Yellow-browed Warbler***, near Colyton on 30th.

November

Two **Dippers** were on the Coly on 1st, with one here again a week later, and the 2nd showed another large movement of **Woodpigeons** overhead. Eight **Brent Geese** flew west at sea on 4th, and 6th gave the passerine of the month, with a **Serin** east over Haven Cliff. A nice wildfowl movement on 7th included six **Brent Geese**, eight **Red-breasted Mergansers**, one **Eider**, three **Tufted Ducks** and three **Shoveler**, with a single **Storm Petrel** being an unseasonal surprise. Also on this date a **Knot** appeared on the estuary, remaining for the rest of the week. Duck movement continued over the next few days, with **Red-breasted Mergansers** being surprisingly frequent. A late **House Martin** flew towards Seaton town on 8th, though more expected given the time of year was a **Pochard** which joined the **Tufted Duck** on the Borrow Pit the following day. Also on 9th, two **Short-eared Owls** were found late afternoon over Colyford Marsh. They were seen on and off till at least 13th. The 10th saw a **Little Auk** fly west past Seaton, and the following day the **Velvet Scoter** count rose to four. Four **Black-necked Grebes** were in the bay briefly on 12th, and another three **Little Auks** passed through. A **Pomarine Skua** past Branscombe on 13th was a surprise given the calm conditions, 18 **Brent Geese** also flew through. Two **Firecrests** appeared in Jubilee Garden; sightings of this species here continued into December – though at times they could be incredibly elusive! Singles of **Little Auk** and **Goosander** passed over the sea on 14th and 15 **Golden Plover** beside the A3052 near Seaton Water Tower were obviously a result of the recent cold snap. A **Long-tailed Duck** fed close inshore off Seaton beach, after it had first flown past Branscombe on 17th; also on this date four more **Goosander** flew north over Seaton Marshes. Three **Pintail** and a **Gadwall** were on a flooded Bridge Marsh on 21st and a couple of **Bramblings** were found at Colyford Common, remaining here till at least the month's end. The 23rd saw the reappearance of a (or another) **Short-eared Owl** over Colyford Marsh for one evening only and 24th saw yet another duck highlight for the month with a female **Mandarin** on the Borrow Pit. Other highlights on this date included a **Jack Snipe** on Axmouth Marsh and a pair of **Gadwall** in the area. A good seawatch on 25th saw another **Little Auk** fly west, and birds on the sea included six **Great Crested Grebes**, two **Red-throated Divers** and an increase to five **Velvet Scoters** (though the total had dropped back to four by the following morning). Five **Woodcock** were back in their usual wood near Colyton on 29th, and 30th saw yet another **Little Auk** pass Seaton.

December

A male **Black Redstart** on Beer Beach on 1st remained elusive here throughout the month. There were four more on patch by the end of the month – all female-types – with two at the Yacht Club, one in Durley Road and another in Beer. A **Greylag Goose** flew upriver on 3rd, and 5th produced a surprise **Pomarine Skua** off Seaton, along with six passing **Brent Geese**. A **Spoonbill** spent a few hours on the estuary on 14th, and 15th saw an increase in **Velvet Scoters** with eight in the bay, though only briefly. A good day on 16th produced two **White-fronted Geese*** on Colyford Marsh which remained till Boxing Day and a **Grey Partridge** at Rousdon which was seen on/off till the year's end. An adult **Yellow-legged Gull*** was on the estuary on 17th and four **Great Northern Divers** flew east the following morning, with a **Short-eared Owl** over Colyford Marsh that evening. **Golden Plover** started appearing in small numbers with the cold, which probably also prompted the appearance of the odd duck, with four **Gadwall** at Colyford on 19th, a **Tufted Duck** at Lower Bruckland Ponds on 21st, a **Pintail** here on 24th, with seven of this species on the sea off Seaton the next day (including five drakes). Another **Yellow-legged Gull***, a fourth-winter, put in a few appearances on 21st and 23rd and **Mediterranean Gulls** became more evident with approximately ten individuals through during the Christmas week.

The Axe Estuary and Seaton Bay Bird Report 2007

Introduction to Species Accounts

The following notes are provided to facilitate the use and interpretation of the individual species accounts.

Sources
The vast majority of records have been extracted from the personal notes of several local birdwatchers, and from the logbooks at Seaton Marshes and Colyford Common Local Nature Reserves.

Species
All species in the systematic list are in Categories A, B and C of the British List maintained by the British Ornithologists' Union (BOU). The nomenclature adopted is that generally used in the Devon Bird Report.

Status
A comment on status is included for each species and subspecies, and refers to its occurrence within the recording area. This is intended to give the reader a reasonably clear idea of the relative rarity or abundance of that species, whether it breeds and, in many cases, at what time of year it is likely to occur. In order to give some consistency to these comments, the following guidelines have been applied:

Status	Number of individuals or records annually
Rare	<3
Scarce	3 - 20
Fairly Scarce	21 - 100
Fairly Common	101 - 500
Common	500+

There is some flexibility in their application, however, so that a large and obvious species like Mute Swan, whilst never numbering over 100 individuals, could easily give rise to 100+ records, and is noted as 'fairly common', whilst another highly visible species, Peregrine, may easily produce 100+ records but is numerically far less abundant than Mute Swan, so is noted as 'fairly scarce'. Whilst compiling our species list it became clear that there are some gaps in our knowledge of the local avifauna, so that we cannot even say with certainty whether certain species breed, for example. Also, increased watching is likely to 'downgrade' some species from 'rare' to 'scarce', and so on. In summary then, the status comments are no more than a rough guide, based on often incomplete data.

The Axe Estuary and Seaton Bay Bird Report 2007

Rarity Protocols

The Systematic List includes both national and county rarities. The former are assessed by the British Birds Rarities Committee (BBRC), the latter by the Devon Birds Records Committee (DBRC). In order to make this report as complete an account of the year's birds as possible, we have decided to include all rarity records, as long as they have been submitted to the relevant body. The species account will make clear whether the record has been accepted, or is still pending. Any pending records that are subsequently rejected will be notified in next year's report. Obviously, rejected records are not included!

Abbreviations

These have been kept to a minimum, but for brevity some have been used. Most will be self-evident, but in the case of birds recorded passing (especially on seawatches), something like the following may be seen: 'eight W, 36E' – meaning eight flew west, and 36 flew east. Also, WTW = Waste Water Treatment Works, LNR = Local Nature Reserve, NNR = National Nature Reserve.

Tables

Often the term 'monthly maximum' is used, indicating the highest single count made during each month.
Occasionally the term 'bird-days' is used. It represents the sum total of birds recorded each day during the period in question (usually one month). Bird-day figures should be interpreted with caution. For example, '16 bird-days' could indicate a) 16 different single birds on 16 dates, b) the same two birds on eight dates, perhaps consecutive, c) a flock of 16 on one day, or a whole host of other permutations!

Species List

At the end of the report is a list of species that are known to have occurred in the recording area. Where a species is a national or county rarity, criteria for inclusion will normally include acceptance of the record by the BBRC or DBRC. However, a few records predate the formation of these bodies, but are nonetheless considered to be reliable by them (e.g., Little Bittern) and are included in the area list.

Update to 2006 Species Accounts

In the 2006 account several records were notified as 'pending consideration by BBRC/DBRC'. All records were subsequently accepted, apart from the following:
Rejected: Great White Egret, Barred Warbler, Nightingale.
Still pending: Gull-billed Tern.
Unfortunately the loss of Barred Warbler means that the recording area species list was reduced by one, though hopefully not for long!
Finally, Goldeneye was omitted in error – the single record has been included in this year's account.

The Axe Estuary and Seaton Bay Bird Report 2007

Species Accounts

Mute Swan - Peregrine	Ian McLean
Water Rail - Waders	Gavin Haig
Skuas - Nightjar	Steve Waite
Swifts - Hirundines	Bob Olliver
Pipits - Wagtails	Steve Waite
Dipper - Firecrest	Donald Campbell
Flycatchers - Starling	Bob Olliver
Sparrows - Buntings	Steve Waite

Mute Swan *Cygnus olor*
Fairly common resident breeder and winter visitor
Present throughout the year on both reserves and neighbouring fields in the Axe valley. Mostly 15 birds or fewer, but numbers were generally slightly up from last year in the first winter period but lower in the second winter period. In March there was an influx that resulted in a peak count of 29 towards the end of the month. Two yellow-ringed individuals arrived at this time, VPB which was ringed at Abbotsbury in 2004 and VZB which was ringed at Radipole in 2005. The only breeding record was a pair with seven cygnets seen on the river near Bridge Marsh.

Whooper Swan *Cygnus cygnus*
Rare winter visitor
Two records, which constitutes a very good year for this less than annual visitor.
One, a second-winter bird, joined the Mute Swan flock just north of our recording area between Whitford and Kilmington on December 23rd 2006 and stayed until the April 10th 2007. Another was seen flying north over Lower Bruckland Ponds on Oct 15th.

Black Swan *Cygnus atratus*
Rare escape/feral visitor
A first-year was seen at Boshill on the morning of Mar 1st and later on the River Axe. (Not included in species totals).

White-fronted Goose *Anser albifrons*
Rare winter visitor
A. a. albifrons (**'European' White-fronted Goose**)
Two adults were found at Colyford Marsh on Dec 16th and stayed until 26th. This species has now occurred in the recording area for the last four years.
Pending consideration by DBRC

Greylag Goose *Anser anser*
Rare winter/feral visitor

A good year for this species as there were five sightings involving a total of eight birds.
One flew in off the sea at 16:20 hrs on Feb 5th and landed at Lower Brucklands Ponds. The bird stayed around the estuary until Feb 21st, generally on Bridge Marsh. The same or another individual was seen near Boshill on Mar 1st. There was one at Colyford Marsh during the morning of Apr 18th. Four arrived at Colyford Marsh on May 3rd and stayed until May 7th when they were seen to fly off towards Shute Hill. Two re-appeared on May 10th but by 17th there was only one bird present, which stayed until May 31st.
One flew up-river at dusk on Dec 3rd.
It is tempting to speculate that one or more of the sightings might have involved truly wild birds (they were very wary and stayed away from people) but the odds are slim, although some Scandinavian birds do winter in Spain.

Canada Goose *Branta canadensis*
Fairly common naturalised introduction, breeds

Present throughout the year with a maximum of 30 in November; generally numbers were similar to 2006. Usually encountered on the estuary marshes or at Lower Bruckland Ponds. The only breeding record was of a pair with a single gosling at Seaton Marshes.

Barnacle Goose *Branta leucopsis*
Rare winter/feral visitor/escape

An average year with a two sightings involving a total of three birds.
Two arrived at Colyford Marsh on May 3rd, the same day as four Greylag Geese, but while the Greylags started an extended stay on the estuary the two Barnacles left the next day. They were wary and stayed well away from people, so it is possible that they might have been wild birds but the odds are slim due to the increasingly large number of feral birds in the country.
There was one on the scrape at Colyford Marsh on the morning of Jun 12th.

Barnacle Geese *by Roger Boswell*

Brent Goose *Branta bernicla*
B. b. bernicla (Dark-bellied Brent Goose)
Scarce winter and passage visitor

A very good year with a total of 200 birds seen; the vast majority were fly-bys. A few did linger awhile on the estuary. The first of the year involved a lone bird on the sea on Jan 9th, while the last were five W on Dec 8th. Highest count was a flock of 60 E on Feb 18th.

B. b. hrota (Pale-bellied Brent Goose)
Rare/scarce passage visitor
On Mar 11th five drifted W on the sea with one dark-bellied bird. One on Jun 8th (which arrived in the company of Canada Geese) then wandered around the estuary until Jun 21st. This was almost certainly the long-staying tame individual that had spent the last year at Abbotsbury just having a fortnight's holiday at sunny Seaton.

Egyptian Goose *Branta ruficollis*
Rare feral visitor/escape
One remained from 2006 and wandered around the estuary until Apr 8th when it was joined by a second bird. Although they were seen copulating on a number of occasions they are not thought to have nested. They were last seen on Jun 18th. One reappeared on Sep 7th and stayed into 2008.

Egyptian Goose by Gavin Haig

Shelduck *Tadorna tadorna*
Fairly common migrant breeder and winter visitor
Numbers were high at the beginning of the year with a record 139 seen in January. A pair of Shelduck was seen on Jun 12th bringing nine small ducklings from Beer to the Axe Estuary. They probably nested on the cliffs E of Branscombe. Numbers of young seemed quite good this year on the estuary where they formed creches, with a couple of adults taking care of many ducklings.

2007	Jan	Feb	Mar	Apr	May	Jun	Jul	Aug	Sep	Oct	Nov	Dec
Maxima	139	107	104	54	95	85	55	8	5	5	26	54

Shelduck ducklings by Roger Boswell

The Axe Estuary and Seaton Bay Bird Report 2007

Mandarin Duck *Aix galericulata*
Rare naturalised introduction/escape
Sightings of this species are becoming annual (recorded for the last three years), probably due to the increasing feral population. A female spent the morning of Nov 24th on the Borrow Pit at Seaton Marshes.

Wigeon *Anas penelope*
Common winter visitor
Counts were back to normal after last year's record numbers (1200 in February 2006). Although Colyford Marsh is probably the favoured location, birds may be spread widely through the estuary.

2007	Jan	Feb	Mar	Apr	May	Jun	Jul	Aug	Sep	Oct	Nov	Dec
Maxima	550	268	145					2	42	187	426	400

Gadwall *Anas strepera*
Scarce passage and winter visitor
An average year with eight sightings involving a total of 12 birds, three in the first winter/spring period and five in the autumn.

Teal *Anas crecca*
Fairly common passage and winter visitor
Nothing out of the ordinary this year, though good numbers during autumn passage. Teal can often be scattered widely on the estuary and marshes and can be elusive, so are probably under-recorded.

2007	Jan	Feb	Mar	Apr	May	Jun	Jul	Aug	Sep	Oct	Nov	Dec
Maxima	92	34	110	8				40	40	70	130	100

Mallard *Anas platyrhynchos*
Common resident breeder and winter visitor
Present throughout the year in good numbers (often 100+). However, it should be noted that not all are genetically pure, and many are bred and released locally for shooting purposes! Therefore, unravelling the true status of Mallard in our area is nigh on impossible.

Pintail *Anas acuta*
Scarce winter visitor
An average year with sightings involving a total of 18 birds, two in the first winter period and six in the autumn/second winter period. Peak count was a group of seven (five drakes) in Seaton Bay on Dec 25th.

Garganey *Anas querquedula*
Rare passage visitor
A second good year for this species as there were three records, two in spring and one in autumn.
The first was a male at Seaton Marshes on Apr 28th, which was seen later the same day flying downriver and at Colyford Marsh. Another male spent most of May 3rd on the Borrow Pit at Seaton Marshes, seen later that day on the river opposite Colyford Common. Finally, an adult female was at at Colyford Common late in the evening of Jul 20th.

Garganey *by Steve Waite*

Shoveler *Anas clypeata*
Fairly common winter visitor
First winter period counts were slightly lower than usual.

2007	Jan	Feb	Mar	Apr	May	Jun	Jul	Aug	Sep	Oct	Nov	Dec
Maxima	25	18	8		2	2	7	9	1	4	24	18

Pochard *Aythya ferina*
Scarce winter visitor
An average year with three records. There was a male at Lower Bruckland Ponds on Jan 2nd, a female on the scrape at Colyford Marsh on the evening of May 30th and a female at the Borrow Pit, Seaton Marshes from Nov 9th to 17th.

Tufted Duck *Aythya fuligula*
Rare/scarce winter visitor
A good year with eight sightings involving a total of ten birds. One spring record, one in summer and six in the autumn/winter. Unusual were some records from the sea: two were with the Common Scoter flock on Oct 31st and one on Nov 6th, plus three W on Nov 7th.

Eider *Somateria mollissima*
Scarce winter and passage visitor
A relatively poor year with just three sightings involving a total of 25 birds. First, a flock of 22W past Branscombe on Apr 22nd included nine adult males; an immature male W past Branscombe on Jul 1st; two female types E past Seaton on Nov 7th.

Long-tailed Duck *Clangula hyemalis*
Rare/scarce winter visitor
The first since January 2005 was a first-year bird initially seen taking off from the sea at Branscombe on Nov 17th and flying east in the company of a Common Scoter. It was relocated just off Seaton Hole and spent the morning feeding in Seaton Bay.

Long-tailed Duck by *Steve Waite*

Common Scoter *Melanitta nigra*
Fairly common passage and winter visitor
An average year with no exceptional numbers either on passage or wintering.
Winter: In the first winter period numbers were generally in single figures but there were 19 on Jan 26th. Higher numbers in the second winter period included counts of 30 on Dec 10th and 15th.
Passage: Main spring passage was between the end of March and mid-June, with birds seen regularly throughout the period. The best day was May 11th with 41E and 13W. Autumn passage was from the beginning of July to the beginning of November with significant numbers as follows: 40W on Jul 8th, 10E and 32W on Sep 20th and seven E and 30W on Nov 7th.

Surf Scoter *Melanitta perspicillata*
Rare vagrant from North America
The first-winter male off Beer from Dec 19th 2006 stayed until Apr 19th. It could usually be found off the eastern end of Beer beach but on Jan 4th it made a brief visit to Branscombe. By the end of its stay it had begun to develop the white nape patch. This was the first record for the recording area, although two were just outside the boundary, at Weston, from January to April 2005.
Accepted by DBRC

Velvet Scoter *Melanitta fusca*
Rare/scarce winter and passage visitor
An excellent year with record numbers being seen, including some long-staying individuals. A minimum of ten birds was involved.
The first was one W on Feb 24th. A juvenile was on the sea off Seaton Hole on Oct 17th. It remained in Seaton Bay and was joined by a second first-year bird on Oct 25th. They both remained in Seaton Bay until Nov 11th when they were joined by two more first-year birds. The flock increased to five on Nov 25th but, although five birds were present early morning on 26th, by the afternoon they were back down to four. Numbers were boosted to a record eight on Dec 15th but they did not stay, and all flew W at around midday. The following day we were back to four but this fell to three by 18th and these remained until the end of the year.

Goldeneye *Bucephala clangula*
Rare/scarce winter and passage visitor
A single sighting was very welcome for this less than annual visitor. There was a female on the River Axe adjacent to Axe Marsh on the unusual date of Jun 27th.
2006 Record: a male W past Branscombe on Nov 3rd.

Red-breasted Merganser *Mergus serrator*
Rare/scarce winter and passage visitor
An excellent series of records with ten sightings involving a total of 24 birds, a local record. All were fly-bys seen on seawatches between Oct 28th and Nov 26th except for two on the sea off Seaton on Dec 17th. Peak count was eight W on Nov 7th.

Goosander *Mergus merganser*
Rare winter and passage visitor
After none in 2006, eight sightings involving a total of ten birds was exceptional and again is a local record.
The first was a male on the River Axe on Feb 17th. On Apr 13th a female flew over Seaton Marshes and landed on the river near Coronation Corner. There was a juvenile on the river on Sep 11th and 12th. On Sep 17th a female type flew east over Seaton and then north up the River Axe, with possibly the same bird flying downriver past Colyford Common later. On Nov 14th two, a male and a female, flew west past Seaton. On Nov 17th four, three males and a female, flew over Seaton Marshes at 15:15 and headed north up the valley.

Red-legged Partridge *Alectoris rufa*
Fairly scarce and probably breeds, but many reared for release
Records this year included one on the tramway at Seaton Marshes on Mar 26th, and up to four at Beer Head in March and April. Farmland between Axmouth and Rousdon is the best local area for this species – presumably as a result of releases – with a maximum of 20 there in December.

Grey Partridge *Perdix perdix*
Rare, with any records probably the result of birds reared for release
There were two at Axmouth on Apr 16th, two at Haven Cliff on Apr 18th, and one with 20 Red-legged Partridges in December between Axmouth and Rousdon.

Pheasant *Phasianus colchicus*
Common resident breeder, with large numbers reared for release
Encountered throughout the recording area, even in the towns and villages.

Red-throated Diver *Gavia stellata*
Fairly common winter and passage visitor
Numbers were well down on the last couple of years, especially in the first winter period.

2007	Jan	Feb	Mar	Apr	May	Jun	Jul	Aug	Sep	Oct	Nov	Dec
Maxima	17	7	2	1	2	1			1		5	2

Winter periods: The reduced numbers in the first winter period were probably due to the beaching of the MSC Napoli on our most productive area for this species – a reef off Branscombe (see article on p.86). The peak in January was before the incident, following which a number of Red-throated Divers were seen to be oiled. There was only one record of more than ten birds, on Jan 10th, when there were four on the sea off Branscombe and thirteen flew W. Average numbers for the second winter period.

Spring Passage: A steady spring passage took place between Mar 23rd and May 19th, with eleven records involving a total of thirteen birds.

Autumn passage: The first was on the very early date of Sep 20th, and was still in full breeding plumage. A total of 21 birds passed on nine days between Sep 20th and Nov 28th, with a maximum of five on Nov 25th.

Black-throated Diver *Gavia arctica*
Scarce winter and passage visitor
Another very good year for this species. There were eight sightings involving a total of nine birds, including two seawatching records, as follows: singles off Seaton on Jan 3rd, Branscombe on Jan 4th, Beer on Jan 10th, and Seaton on January 26th and 27th; two off Branscombe on March 13th, and singles W on May 11th and 31st.

Great Northern Diver *Gavia immer*
Fairly scarce winter and passage visitor
A very good year with more seen in both the first winter period and on spring passage than in 2006.

2007	Jan	Feb	Mar	Apr	May	Jun	Jul	Aug	Sep	Oct	Nov	Dec
Maxima	1	2	1	1	11	1			1	1	4	5
Bird-days	10	5	1	1	62	1			2	1	20	16

The **spring passage** was excellent with a total of 63 birds (31 in 2006) seen on fourteen days between May 8th and Jun 2nd. Most were seen on seawatches, flying west. Peak count was 11W on May 11th. Seawatching was quite popular during this period due to the presence of Storm Petrels in Seaton Bay.

Autumn passage occurred between Sep 20th and Nov 28th when a total of 23 birds passed on 14 days with a maximum of four W on Nov 13th.

Unidentified Divers *Gavia sp*
Distant divers in flight can be a real identification challenge, especially Black-throated Diver. A total of 12 birds went unresolved by experienced seawatchers this year, mostly during spring passage.

Little Grebe *Tachybaptus ruficollis*
Fairly scarce resident breeder and winter visitor
Although present on the Borrow Pit at Seaton Marshes throughout the breeding season no young were seen. Noted throughout the year in small numbers, mainly on the River Axe, with a maximum of eight in March and November.

Great Crested Grebe *Podiceps cristatus*
Fairly scarce winter and passage visitor
May be encountered anywhere off the coast in the winter months, but seems to have a preference for Seaton seafront. Monthly maxima are given in the following table.

2007	Jan	Feb	Mar	Apr	May	Jun	Jul	Aug	Sep	Oct	Nov	Dec
Maxima	4	5		1	1		2			2	6	4

Slavonian Grebe *Podiceps auritus*
Rare winter and passage visitor
A below average year with only a single sighting of two together off Seaton seafront on Jan 27th.

Black-necked Grebe *Podiceps nigricollis*
Rare winter and passage visitor
A better than average year with two records involving a total of five birds.
On Jan 4th one was close in off Beer but unfortunately only stayed five minutes before flying off west, also being seen flying along the line of the surf at Branscombe. On Nov 12th there were four on the sea off Seaton, but again they did not linger.

Fulmar *Fulmarus glacialis*
Fairly common passage visitor; small numbers breed
A cliff nesting survey carried out in mid-April found a total of 22 nests between Seaton Hole and Branscombe. Birds were present from the beginning of the year until the end of September, and after an absence of just over six weeks birds had returned by mid-November.

There was an unusual record of one about a mile inland circling over Seaton on Apr 30th. Good numbers were occasionally seen during seawatches, including 100+W on Jan 18th, 40W on May 10th, 42W on May 11th and 43W and one E on May 12th.

Sooty Shearwater *Puffinus griseus*
Rare/scarce autumn passage visitor
A record year for this species both in terms of peak count (seven on Sep 21st) and a record total of ten for the year. In addition to the birds of Sep 21st there was one on Sep 16th and two on Sep 20th. Most flew W. (Note that on Sep 21st a total of 26 birds were logged as 'unidentified shearwater sp'. It is highly likely that some of these also were Sooty Shearwaters).

Manx Shearwater *Puffinus puffinus*
Common passage and summer visitor, most numerous in spring.
Good numbers were again seen in May and June, though they did not match last year's record counts.
Recorded between April 1st and September 19th, the following table gives the monthly maxima.

2007	Jan	Feb	Mar	Apr	May	Jun	Jul	Aug	Sep	Oct	Nov	Dec
Maxima				49	252	156	137		2			

There were only four counts of over 100 birds: 252 on May 7th, 156 on Jun 20th, 132 on Jun 21st and 137 on Jul 5th.

Balearic Shearwater *Puffinus mauretanicus*
Fairly scarce passage visitor, mainly in late summer and autumn.
A second excellent year with a total of 128 birds recorded between Jun 14th and Oct 28th, but not quite up to last year's record of 137. There was some compensation – a record day total of 64W on Sep 20th.

2007	Jan	Feb	Mar	Apr	May	Jun	Jul	Aug	Sep	Oct	Nov	Dec
Maxima						4	6	10	64	3		
Bird-days						11	14	17	75	3		

Other double figure counts were: ten W on Aug 18th and ten W on Sep 21st.

The Axe Estuary and Seaton Bay Bird Report 2007

Unidentified Shearwaters *Puffinus sp*
Distant or briefly seen shearwaters can be difficult to positively identify to species and even experienced seawatchers have to leave some birds unresolved.
A total of 40 birds seen between Aug 18th and Sep 23rd were logged as 'shearwater sp.', with 26 of those seen on Sep 21st, some of which were thought to be Sooty Shearwaters.

Storm Petrel *Hydrobates pelagicus*
Normally rare passage visitor, though prone to influxes
After last years storm driven influx which produced record numbers it was a pleasant surprise to find that Storm Petrels again chose to visit our part of the coast in late spring in good numbers. A total of 87 was seen compared with 211 in 2006 and only one the previous year. While the numbers were generally lower this year, they occurred over a much longer time period – two months – compared with just ten days in 2006.
A total of 86 birds was seen between May 10th and Jul 13th, with peak counts of ten with fishing boats on Jun 25th, and 11W on Jul 6th. Also, a single autumn record of one behind a fishing boat on Nov 7th.

Storm Petrels were also seen on the two pelagics out of Beer with 116 seen on May 16th and 40 on June 27th. These involved an observer or two going out with the Beer fishermen as they did the rounds of their lobster pots travelling up to four miles out. As the recording area is officially within a 5km radius from the old Axe Bridge these counts were not included in the overall totals.

Gannet *Morus bassanus*
Common passage visitor
An average year, with birds being seen in all months, and a noticeable increase in rough weather. Significant numbers (>100) as follows:
Three E and 150W on Jan 18th, 17E and 112W on May 11th, 143E and eight W on May 27th, 173W on Jun 12th, 113W on Jun 21st and 199W on Dec 2nd.
In common with most seabirds in windy conditions the majority of birds flew into the prevailing wind.

Cormorant
Phalacrocorax carbo
Fairly common resident breeder and winter visitor
A cliff nesting survey carried out in April, between Seaton Hole and Branscombe, found a total of 37 nests. They are to be seen regularly along the River Axe and along the coast.

Cormorant *by Roger Boswell*

Shag
Phalacrocorax aristotelis
Fairly scarce resident breeder and winter visitor
Although birds were present during the breeding season (for example, 11 seen from Beer Head on Apr 8th) no nest sites were found during the cliff nesting survey, but it should be noted that parts of the cliffs around Beer Head are impossible to survey from the land.

Cattle Egret *Bubulcus ibis*
Rare vagrant
One was found in the egret roost at Seaton Marshes on Oct 14th and seen around the estuary marshes on and off over the next few days. It was very elusive during the day, however, and its main feeding area was never discovered. It was last seen flying W past Beer Head at 09:08 on Oct 21st. Probably the second record for the area, following one at Colyford Common in July 2005.
Accepted by BBRC

Cattle Egret *by Gavin Haig*

Little Egret *Egretta garzetta*
Fairly common resident breeder and winter visitor
Three or four pairs attempted to nest in the heronry above Axmouth but only one pair was successful, with two juveniles seen in the nest and subsequently on the river. This is the first confirmed breeding record for the Axe Estuary. The largest numbers were again seen during the winter period with a maximum of 39 birds on Dec 10th.

Juvenile Little Egret *by Roger Boswell*

Grey Heron *Ardea cinerea*
Fairly common resident breeder
Seen throughout the year. Four or five pairs nested in the heronry above Axmouth and appeared to be reasonably successful with good numbers of juveniles seen on the estuary after the breeding season.

Black Stork *Ciconia nigra*
Rare vagrant from Europe
The following are the first records for the recording area.
One flew out of a stream at Southleigh at 05:45 on Jun 8th and headed towards Northleigh, but was not relocated. Discussions with local people suggested that it had been present since Jun 4th and was last seen on Jun 9th, but avoided being seen by the majority of local birders.
On Aug 6th one flew low NW over Seaton.
Both sightings pending consideration by BBRC

Spoonbill
Platalea leucorodia
Scarce passage visitor
After last year's record numbers it was a little disappointing that there was only one sighting this year. A first-year bird spent the afternoon of Dec 14th on the estuary. It was found at Coronation Corner and last seen near the old Axe Bridge.

Juvenile Spoonbill *by Karen Woolley*

Honey Buzzard *Pernis apivorus*
Rare passage visitor.
The following three sightings constitute the first for the recording area. At 07:25 on Jun 18th one flew south over Colyford Common, rested briefly in a tree before flying off to the north. It was thought to be probably a male. One, also a probable male, was photographed flying north over Colyton on Aug 3rd. Finally, another was photographed over Haven Cliff on Sep 2nd.
All records accepted by DBRC

Red Kite *Milvus milvus*
Rare/scarce passage visitor.
Four records this year was perhaps average, and compares with five in 2006. As the British breeding population of this spectacular raptor increases we might expect further increases in sightings.
The first was one that was seen over Beer for five minutes at 10:10 on March 16th before flying off slowly to the east. The following day one came in off the sea and then flew north-east over Axe Cliff Golf Course. One was seen flying northwest over Seaton at 15:30 on April 6th and finally one was seen flying north from Colyford Common on May 24th.

Marsh Harrier *Circus aeruginosus*
Rare/scarce passage visitor.
An average year, with just a single sighting; a little disappointing after last year's record six. An adult male hunted over the reed bed at Colyford Common late in the afternoon of Sep 29th.

Hen Harrier *Circus cyaneus*
Rare passage visitor.
Two records in 2007 were the first for several years. A male briefly over Colyford Marsh on Jan 4th, and then in autumn a 'ringtail' flew north over Haven Cliff on Oct 31st.

Goshawk *Accipter gentilis*
Rare/scarce passage and winter visitor.
An average year with two records – singles in March and September

Sparrowhawk *Accipiter nisus*
Fairly common passage visitor and resident breeder.
Breeds locally and seen regularly throughout the recording area.

Buzzard *Buteo buteo*
Common passage visitor and resident breeder.
Breeds locally with regular sightings over the estuary and surrounding hills. The maximum number seen was 20 from Colyford Common on Apr 18th.

Buzzard by Roger Boswell   **Osprey** by Roger Boswell

Osprey *Pandion haliaetus*
Scarce passage visitor

Another excellent year for this species with a probable total of nine birds – one spring record and eight in autumn (a record number for the area). While all of last year's birds passed through quite rapidly, a long staying juvenile in autumn delighted both local and visiting birders.

The spring bird flew N over Seaton Marshes on Apr 2nd.

The first in autumn was on the morning of Jul 24th and lingered long enough to catch a fish from the river before flying off W. Next was one over the river late in the afternoon on Aug 10th. On Aug 17th two were seen, the first in the morning over Seaton Bay before drifting W and the other in the afternoon, flying SE over Seaton and out to sea. On Aug 30th one was seen flying N over the seafront and up the river. A juvenile arrived on Sep 7th and stayed until Sep 17th; on that date it was seen circling over Haven Cliff, gaining height until it was a small dot in the sky before drifting S. During its stay it roosted to the north of the A3052 and then came down the river to fish. On Sep 16th a second bird was seen and was thought to have passed through quickly.

Kestrel *Falco tinnunculus*
Fairly common passage visitor and resident breeder
Breeds locally with regular sightings over the estuary. The maximum together was four at Haven Cliff on Sep 10th; this included two juvenile birds.

Kestrel *by Fraser Rush*

Merlin *Falco columbarius*
Scarce passage and winter visitor
A good year with a total of nine records; three in winter, one in spring and five in autumn.
One at Musbury on Jan 9th. A female at Colyford Common on Feb 1st, and another there on Feb 20th.
One in spring over Seaton Marshes on Apr 2nd.
The first autumn record was one at Beer Head on Sep 29th. A female flew E at Haven Cliff on Oct 3rd, and another female was there on Oct 29th. On Oct 31st a juvenile male flew W over Coronation Corner. The last of the year was one E past the farm gate viewpoint on Nov 15th.

Hobby *Falco subbuteo*
Fairly scarce passage and summer visitor.
A good year, with the first being seen on Apr 23rd and then fairly regularly until Sep 25th. Numbers were similar to last year except for a more than doubling of the number of sightings in July, which probably indicates that they are breeding somewhere nearby. Monthly maxima and bird-day totals are shown in the following table.

2007	Jan	Feb	Mar	Apr	May	Jun	Jul	Aug	Sep	Oct	Nov	Dec
Maxima				1	1	2	2	2	1			
Bird-days				3	5	7	16	5	6			

Peregrine *Falco peregrinus*
Fairly common resident breeder and winter visitor
Breeds locally with regular sightings over the estuary. They were not thought to have bred very successfully this year as only a single juvenile was seen after the breeding season.

Water Rail *Rallus aquaticus*
Fairly scarce winter visitor and resident breeder
Breeding again proven this year, with single juveniles seen in June and August, plus two small chicks with an adult, also in August, suggesting at least two successful attempts. No counts greater than four during the year, though no doubt many more than that were present during the winter months.

Corncrake *Crex crex*
Rare passage visitor
One flushed during ringing operations at Colyford Common during the afternoon of Sep 24th appears to be the first area record since 1929!
Accepted by DBRC

Moorhen *Gallinula chloropus*
Fairly common resident breeder
Breeding records this year included a brood of five on a tiny pond in Beer Quarry. Otherwise, maxima included 21 at Lower Bruckland Ponds in February, 18 at Seaton Marshes in October and 15 at Colyford Common in November.

Coot *Fulica atra*
Fairly scarce resident breeder
At least three nesting attempts at Lower Bruckland Ponds produced four or five chicks, and the highest count of adult birds was six here in February. Elsewhere, recorded at Seaton Marshes, with up to four.

Oystercatcher *Haematopus ostralegus*
Fairly common passage and winter visitor, fairly scarce in summer
A typical showing, with double figures, but fewer than 20, in each winter period. A mating pair were seen on Beer Beach on Apr 10th, but there was no subsequent evidence of breeding. Occasionally recorded on seawatches, usually ones and twos, though six W on May 10th. The table lists counts made on the estuary and marshes.

2007	Jan	Feb	Mar	Apr	May	Jun	Jul	Aug	Sep	Oct	Nov	Dec
Maxima	12	7	7	4	6	2	7	2	8	8	12	17

Stone-curlew *Burhinus oedicnemus*
Rare passage visitor
One frequented Seaton Marshes and, later, the estuary on Apr 13th and was the first local record for 44 years! The bird was colour-ringed and originated from the Breckland population, where it had hatched in 2005.
Accepted by DBRC

Stone-curlew (note colour-rings) *by Karen Woolley*

Little Ringed Plover *Charadrius dubius*
Scarce passage visitor.
The Axe estuary is one of the best localities in Devon to see this species
A lighter passage than 2006, with a minimum of seven birds (15 last year), all but one in spring. The first was at Seaton Marshes on Mar 26th, with another there on Apr 2nd. On Apr 13th a single flew N over Seaton Marshes and may have been one of three found on the Axe estuary later the same day, with one of these, or another, at Colyford Marsh on 14th. The last bird of the spring was seen from the seafront as it flew S from the river early in the morning, and away high to the west.
In autumn, one on the Colyford Marsh scrape from Jul 9th – 11th was the only record.

Ringed Plover *Charadrius hiaticula*
Fairly scarce passage and winter visitor

No winter records this year, but several in spring, and a very good autumn passage The first were four on the estuary on Apr 14th, then single figure counts until an exceptional spring peak of 28 on May 28th. Autumn records began with five on the estuary on Aug 6th, and numbers reached double figures on several occasions, with peaks of 19 on the beach on Aug 15th, 33 on the estuary on 29th, and 22 there on Sep 8th. This compares with a 2006 peak of 13. The scrape at Colyford Marsh was in good condition throughout the autumn, and counts there reached 15 on Aug 27th. Of note, one on the beach on Aug 30th showed characters of the small, dark *tundrae* race.

2007	Jan	Feb	Mar	Apr	May	Jun	Jul	Aug	Sep	Oct	Nov	Dec
Maxima				6	28	5		33	14	4		

Ringed Plover by *Steve Waite*

Golden Plover *Pluvialis apricaria*
Fairly scarce passage and winter visitor, most numerous in very cold weather
One on the estuary with Lapwings on Jan 2nd was the only record until autumn, when one flew W over Beer Head on Sep 10th. Following two at Colyford Marsh on Sep 28th, October proved to be the best month, with migrants over on eight dates, mostly ones and twos, though seven on 10th. After a blank November, ten were in fields near Hangman's Stone on Dec 19th and four near Boshill Cross on 20th.

Grey Plover *Pluvialis squatarola*
Scarce passage visitor
A spring record – one on the estuary Apr 28th-29th. In autumn, singles recorded on the estuary and Colyford Marsh on Sep 28th and Oct 1st, 6th and 8th – conceivably relating to just one or two birds.

Lapwing *Vanellus vanellus*
Common winter visitor
Peak count in the first winter period was 725 at Colyford Marsh in February. Up to three were on the marshes through April to June, though there was no evidence of breeding. A few passage birds noted through late summer, before numbers began to rise again in Oct, reaching a peak of 638 in late December.

Knot *Calidris canutus*
Scarce passage and winter visitor
A typical year. Probably four or five singles as follows: on the estuary Jan 27th, Feb 3rd and 5th, Aug 8th and Nov 7th-10th; also, at Colyford Marsh Aug 25th-26th. Finally, a seawatching record was unusual – seven W past Seaton on Sep 21st.

Sanderling *Calidris alba*
Scarce passage visitor
With a minimum 48 birds seen in 2007 (70+ in 2006) this species still maintains its rarity on the estuary, with just 3 records (five birds). No records until May, which is easily the best month, with at least 42 birds – all but five flew past during seawatches, with a peak of 15W on 12th. Other records were: one on the estuary on Jun 12th, one W from a boat 3 miles off Seaton on Jun 27th, three S over Haven Cliff on Aug 6th, and one on Seaton Beach with Ringed Plovers and Dunlin on Aug 15th.

Little Stint *Calidris minuta*
Scarce passage visitor

After just three autumn singles in 2006 this year produced a glut, with a very conservative minimum of 16 birds, and an autumn passage of 64 bird-days!
Spring records are unusual, so an adult on the estuary and at Colyford Marsh from May 1st-3rd was welcome.
The autumn presence began with two on the Colyford Marsh scrape on Sep 12th. All subsequent birds favoured the same venue, beginning with five from Sep 24th-26th, then a virtually constant presence from Oct 3rd-22nd, building from just one, then three, then five, peaking at eight (on 9th) and dwindling to a single on the last date. Of four present on Oct 13th, one was trapped and ringed. All autumn birds were juveniles.

Adult Little Stint *by Gavin Haig*

Temminck's Stint *Calidris temminckii*
Rare passage visitor

An obliging juvenile on the scrape at Colyford Marsh from Sep 13th -26th was a very welcome addition to both the autumn wader collection and the recording area list. A rare bird in Devon, this individual was only the 24th for the county.
Accepted by DBRC

Juvenile Temminck's Stint *by Steve Waite*

Pectoral Sandpiper *Calidris melanotos*
Rare vagrant

Most British records of this species involve autumn juveniles, so an adult on the estuary on May 23rd was most unusual, and very welcome! Interestingly, although the the last local records were two in autumn 2003, these were preceded by a spring bird in April 2002.
Accepted by DBRC

Pectoral Sandpiper *by Gavin Haig*

Curlew Sandpiper *Calidris ferruginea*
Scarce passage visitor

Compared with last year's three records (just four birds) 2007 was excellent! Spring records are rare in Devon, so a breeding plumaged adult on the estuary on May 22nd was notable, with the same or another there on 24th.

Autumn passage comprised 77 bird-days, beginning with a moulting adult and two juveniles from Aug 29th to Sep 1st, followed by a virtually constant presence of juveniles until Sep 27th, peaking at six from 15th-17th, and five on 20th and 21st. Finally, another juvenile from Oct 8th-17th. Most records were from Colyford Marsh, with birds also seen on the estuary and occasionally roosting on Seaton Beach.

Curlew Sandpiper *by Steve Waite*

Dunlin *Calidris alpina*
Fairly common passage and winter visitor

The table shows monthly maxima from the estuary and marshes. In addition seawatching revealed a very good spring passage, including an exceptional movement on May 10th when at least 364 were counted: 270W from the seafront, and 94 on the estuary all day (see also Whimbrel). Unusual was a record of 27 heading S over Colyton on Aug 7th.

2007	Jan	Feb	Mar	Apr	May	Jun	Jul	Aug	Sep	Oct	Nov	Dec
Maxima	49	33	7	8	181	4	31	37	84	28	11	34

Dunlin by Gavin Haig

Ruff *Philomachus pugnax*
Scarce passage and winter visitor

Six records of probably nine birds is quite a good showing. One on Colyford Marsh with Lapwings on Jan 26th was the only record until another at Colyford Marsh for two days from Aug 23rd. A juvenile on the estuary from Sep 5th-17th was followed by two at Colyford Marsh Oct 3rd-4th (one remaining until 9th), another there on 13th, and three on 29th.

Jack Snipe *Lymnocryptes minimus*
Scarce winter visitor, but no doubt overlooked
Just two records of this diminutive skulker is very poor – a migrant flushed from a stubble field at Haven Cliff on Oct 8th, and one on Axmouth Marsh on Nov 24th.

Snipe *Gallinago gallinago*
Fairly common winter visitor
The table shows monthly maxima from the estuary marshes but, due to the secretive habits of this species, probably does not reflect the true picture.

2007	Jan	Feb	Mar	Apr	May	Jun	Jul	Aug	Sep	Oct	Nov	Dec
Maxima	8	6	2	12	2		1	30	18	8	48	80

Woodcock *Scolopax rusticola*
Rare/scarce winter visitor, but no doubt overlooked
Most records were from Cottshayne Hill Woods, where up to seven were noted in the first winter period, and five in the second. Also, singles at Holyford Woods in January, and Couchill Woods in February and March.

Black-tailed Godwit
Limosa limosa
Fairly scarce passage and winter visitor, but wintering population increasing

A typical year, with around 20 wintering, plus a smattering of passage birds. Of note were two W during a seawatch on Jan 18th and one W over Seaton on Sep 9th. On Aug 15th a migrant flock of 14 was on the estuary, and seven (perhaps part of this flock) on floodwater N of Boshill Cross, at New House Farm.

Black-tailed Godwit by *Steve Waite*

2007	Jan	Feb	Mar	Apr	May	Jun	Jul	Aug	Sep	Oct	Nov	Dec
Maxima	20	22	4	2	2	5	4	14	11	6	19	13

Bar-tailed Godwit Limosa lapponica
Scarce passage and winter visitor
A notable spring passage this year, with 28 bird-days in April, and 15 in May. This included sea-watching records of four E on Apr 28th, six E on 30th, two W on May 5th and three W on 11th; peak count on the estuary was six on Apr 30th. Autumn was much quieter, with several records of singles in September and October – probably six different birds.

Whimbrel Numenius phaeopus
Fairly common passage visitor, most numerous in spring
The table includes seawatching counts as well as those made on the estuary/marshes. A very strong spring passage this year, notably on Apr 24th with 118W plus six on the estuary, and on May 10th when 133 flew W during the day (see also Dunlin); another good count was 53W on May 12th. The biggest count made on the estuary/marshes during this period was 27 on Apr 23rd. One feeding with horses by stables in Beer on May 3rd was an unusual record.

2007	Jan	Feb	Mar	Apr	May	Jun	Jul	Aug	Sep	Oct	Nov	Dec
Maxima				124	133	4	3	3	1			

Curlew Numenius arquata
Common passage and winter visitor
With an expected wintering population of around 200 a few of this year's counts look suspiciously low, though this is probably due to under-recording rather than a lack of birds. In addition, small numbers were occasionally noted passing along the coast.

2007	Jan	Feb	Mar	Apr	May	Jun	Jul	Aug	Sep	Oct	Nov	Dec
Maxima	135	190	50	13	4	35	100	120	94	75	50	160

Common Sandpiper
Actitis hypoleucos
Fairly common passage and fairly scarce winter visitor
Remarkably similar figures to 2006, though the spring peak of seven on the estuary on May 2nd is a little higher than last year's four. Almost all records were from the estuary and marshes.

Common Sandpiper by Karen Woolley

2007	Jan	Feb	Mar	Apr	May	Jun	Jul	Aug	Sep	Oct	Nov	Dec
Maxima	2	1	1	1	7	2	12	10	5	3	2	3

Green Sandpiper *Tringa ochropus*
Fairly common passage visitor, mostly in late summer; occasionaly winters
A better showing than 2006 (when there were none before Jun 28th), probably aided by the consistently good condition of the scrape at Colyford Marsh. A few of the winter records were from the Coly valley.

2007	Jan	Feb	Mar	Apr	May	Jun	Jul	Aug	Sep	Oct	Nov	Dec
Maxima	1	2	2	2	7	6	11	10	6	6	2	1

Greenshank *Tringa nebularia*
Fairly scarce passage visitor, more regular in autumn
Three spring birds: singles on the estuary on Apr 10th, Colyford Marsh on May 13th, and W past the seafront on May 19th.
Autumn passage began with one on the estuary on July 2nd, followed by a light sprinkling until the last two at Colyford Marsh on Sep 22nd. The highest count was five at Colyford Marsh in mid-September, and all records were on the estuary and marshes except two W over Beer Head on Aug 11th.

Wood Sandpiper *Tringa glareola*
Scarce passage visitor, usually found in late summer
The Axe estuary is one of the best localities in Devon to see this species
This year's records suggest a minimum of nine birds, all at Colyford Marsh. The first was on Aug 1st, then a blank three weeks before another on 23rd, and two from 25th-31st. In September three arrived on 3rd, with one of these departing S at dusk, then two on 5th and a single on 24th. October produced one from 2nd-4th, then a late bird on 17th.

Redshank *Tringa totanus*
Common winter visitor; summers in small numbers; breeds
Bred on the estuary again this year, up to four juveniles being seen in May and June..

2007	Jan	Feb	Mar	Apr	May	Jun	Jul	Aug	Sep	Oct	Nov	Dec
Maxima	84	45	30	6	6	27	23	45	65	70	48	50

Turnstone *Arenaria interpres*
Rare/scarce passage visitor
Seven record of nine birds is probably typical, and again the species mostly avoided the estuary mud!
In spring one was on the Seaton Hole rocks on May 5th, and on May 16th four flew W (viewed from a boat off Seaton), one about a kilometre out and three together at five km.
In autumn, singles were on Seaton Beach on Aug 23rd, on the estuary Sep 5th-6th, heading W during a seawatch Sep 19th, and on the estuary again next day.

Table - Skuas in 2007

Monthly totals	Jan	Feb	Mar	Apr	May	Jun	Jul	Aug	Sep	Oct	Nov	Dec
Arctic Skua				1	44	6	14	13	8			
Pomarine Skua					18	1					1	1
Great Skua			2		1	1	1	13				
skua sp.					4					2		

Pomarine Skua *Stercorarius pomarinus*
Rare/scarce passage visitor
Last year was described as probably the best year on the patch for this species with three records involving six birds, so this years total of 21 birds is truly sensational! Furthermore, many of the birds passed unusually close to the coast.
Peak spring count five on May 27th with four on 31st; first and last spring records were three on May 12th and a single E on Jun 1st.
Singles on Nov 13th and Dec 1st were the only records in the second half of the year.

Arctic Skua *Stercorarius parasiticus*
Fairly scarce passage visitor
As with Pomarine Skua, 2006 was described as 'certainly the best year for sightings of this species off Seaton' with 28 records involving a total of 78 birds. This year produced 39 records relating to 86 birds, with peak counts of five on May 9th, seven on Jul 27th and ten on Aug 14th. First and last dates were Apr 25th and Sep 21st respectively. The lack of records in October (and for all skua sp.) just proves how much the weather plays a part in the occurrence of these oceanic birds on patch.

Great Skua *Catharacta skua*
Scarce passage visitor
One W on Apr 25th then singles past on Apr 28th, and May 11th. One lingering around fishing boats for several days in early June was presumably a loitering non-breeder, whereas one E on July 1st was probably an early autumn migrant. Remainder of the autumn would have been average were it not for an unprecedented hour on Aug 14th when ten flew W! Last record was one on Aug 19th

Mediterranean Gull *Larus melanocephalus*
Fairly scarce, but regular, passage and winter visitor

2007	Jan	Feb	Mar	Apr	May	Jun	Jul	Aug	Sep	Oct	Nov	Dec
Peak counts	3	12	9	2	3	1	1	2	1	1	5	6
Min. monthly totals	9	19	12	3	6	2	8	5	1	3	7	10

An odd year, with few in the autumn, and no first-winters at all in the second winter period, possibly suggesting a poor breeding season. High counts were made during rough weather in late February/early March when large numbers of gulls in general were present, and the first double figure count of this species on patch was made on Feb 28th, with 12 on the estuary during the afternoon. The first juvenile of the autumn was seen on Jul 14th, with only three more over the whole of the next month.
Interesting individuals include an adult with a virtually plain white head present from mid December, and the following colour ringed birds:
WHITE 3C10 – seen on Feb 17th; ringed as an adult in Belgium in May 2005 and reported seven times before our sighting – each time in Cornwall or S Devon.
GREEN AATL – a first-winter on Feb 17th; ringed as a chick NW of Hamburg, Germany in June 2006.
RED 5P5 – a first-winter on Feb 28th; ringed as a chick near Wloclawek, Poland in June 2006

Laughing Gull *Larus atricilla*
Rare vagrant from North America
A first-winter sitting on the sea off Seaton seafront on Feb 28th soon took to the air and was last seen flying W along the beach. A bird of the same age had been present further west in the county since the previous November and disappeared from the River Exe on the afternoon of Feb 27th – it is assumed to be the same bird. This is the second record for the patch, following one in 1976.
Accepted by BBRC

Little Gull *Hydrocoloeus minutus*
Scarce passage and winter visitor
All records involve singles – except for one of two birds – and although no large counts were made the total of eight birds was spread throughout the year with records in seven different months; unusually only a couple were related to rough weather conditions.
A first-winter on the estuary on Jan 1st started the year off well, then an adult W past on Feb 28th, an adult on Seaton Marshes with another S down the estuary at dusk on Mar 4th, an adult on the estuary on Apr 13th, a first-winter here on May 2nd, one W on July 1st and finally a second-winter on Colyford Marsh scrape on Oct 16th.

Bonaparte's Gull *Chroicocephalus philadelphia*
Rare vagrant from North America
A first-summer bird found on the estuary north of Coronation Corner late morning on Apr 30th was watched for only twenty minutes before flying off high north. Thankfully it was relocated on the estuary later in the afternoon, where it remained till nearly dusk when it was flushed by the incoming tide. This sighting constitutes the first record for the area. *Accepted by BBRC.*

Bonaparte's Gull *by Steve Waite*

Black-headed Gull *Chroicocephalus ridibundus*
Common passage and winter visitor, present throughout the year, most abundant in winter
The commonest gull in the winter numbering in the thousands, with counts at the Seaton Hole roost in January revealing over 3,000 birds.
Many wintering birds had left by the first week of March with no more than 40 present through most of the next two months, though numbers soon started picking up with 120 counted on Apr 24th. The first juvenile was seen on Jun 21st. 400 were present by late July and 706 counted on Aug 1st, with over 1,000 present by the end of the month.
Coastal passage was noted, with spring movement starting in Feb with 70W in 30 mins on 13th. The only notable count in autumn was 260W on Jul 7th.

Audouin's Gull *Larus audouinii*
Rare vagrant from Southern Europe
The rarity highlight of the year! An adult/sub-adult was picked out in a large gull flock resting from the rough weather on fields north of Seaton Marshes LNR on Aug 14th. It remained here, albeit elusive, for several hours before being lost from view when the whole flock took to the air. A lucky pair of observers relocated the bird at dusk in Seaton, on the roof of the town's holiday village, but despite lots of effort over the following days it was never seen again. This was the first record for the area, also for Devon, and only the fourth in Britain!
Accepted by BBRC

Audouin's Gull *by Stuart Piner*

Ring-billed Gull
Larus delawarensis
Rare vagrant from North America
Three records involving two individuals was a good showing. A small and slightly oiled second-winter was found resting on the estuary on Feb 19th, mid-afternoon, and remained till dusk; amazingly it was refound more than two weeks later on Mar 3rd, when it was first seen on the estuary, then Seaton Marshes. The second bird was an adult at Coronation Corner from mid-afternoon on Mar 7th, before flying S out to sea at 16:20.
Accepted by DBRC

Ring-billed Gull (left) *by Steve Waite*

Common Gull *Larus canus*
Fairly common passage and winter visitor
A new record set this year showing some good spring passage occurred. The table below shows the maxima for each month:

2007	Jan	Feb	Mar	Apr	May	Jun	Jul	Aug	Sep	Oct	Nov	Dec
Maxima	150	749	410	2	2	1	1	1	4	2	35	70

Late February and early March was a good time for gulls on the Axe, with rough weather coinciding with the peak passage time, this resulting in large numbers present. The count of 749 on the estuary on Feb 28th is most certainly the minimum count for that day, as many were seen heading W at sea during the morning and the true figure was probably well over 1,000 birds.
Interesting individuals include a juvenile on the estuary on Oct 8th, and one still in full juvenile plumage on Oct 28th.

Caspian Gull *Larus cachinnans*
Rare vagrant
The first for the area, and only the second for Devon (following one in Torbay in 2006) occurred on Oct 27th when a second-winter was found on the river during the afternoon. There were large numbers of big gulls on this date, due to the wet and windy conditions, and the collection also included three Yellow-legged Gulls, suggesting an influx from further east.
Accepted by DBRC

Lesser Black-backed Gull *Larus fuscus*
Fairly scarce passage and winter visitor
Usually only small numbers present on the estuary throughout the year (<8), though numbers seem to increase with rough weather, notable counts include: 24 on Feb 12th, 28 on Mar 11th, 25 on May 6th, 34 on Sep 20th and the highest count of the year, 70+ on Oct 28th. Note no three figure counts this year.

L. f. intermedius ('Scandinavian' Lesser Black-backed Gull)
Scarce passage visitor
All records are of adult or near-adult birds.
Only single figure day counts, though a flurry of records in mid-February occurred, with 23 birds (assuming all different birds day to day) between 12th and 21st, and a peak of six on 15th. The only other high count was of five on Mar 15th. The last of the year was one on Oct 28th.
All records of L. f. intermedius pending consideration by DBRC

Yellow-legged Gull *Larus michahellis*
Scarce passage visitor

Another outstanding year for this species, with probably 22 birds, although far fewer juveniles were evident in late summer than in 2006, and birds were seen more sporadically throughout the year. All records relate to birds seen on the estuary unless stated otherwise. Three records in the first half of the year included an adult on Jan 3rd and different single first-winters on Mar 30th and 31st.

In July and August eleven juveniles were seen – all singles – with the first on Jul 16th; this includes one on a flooded field south of Musbury on Aug 14th, and another on Seaton Beach the same morning. The last juveniles were singles on Sept 1st and 4th.

Three (an adult, probable fourth-winter and a second-winter) were on the estuary on Oct 27th, with a fourth bird, a third-winter, on Colyford Marsh the next day. There were two records in December – an adult on 17th and a fourth-winter on 21st and 23rd.

All records pending consideration by DBRC

Juvenile Yellow-legged Gull with two juvenile Herring Gulls (on the right) *by Gavin Haig*

Herring Gull *Larus argentatus*
Common resident breeder and winter visitor
Common all year, breeding throughout the area and regularly present in large numbers on the sea, on the beaches, on the river and in pig fields – no counts were made but several thousand can fly over town at dusk to roost in the bay. There was one record of a colour-ringed individual:
BLUE 102 – an adult on Mar 19th, ringed in November 2005 at Gloucester landfill site near Hempsted, Gloucestershire, and recorded three more times in that county prior to its visit to the Axe.

Iceland Gull *Larus glaucoides*
Rare winter visitor
An amazing series of records in the first winter period, possible relating to as many as six different individuals, all first-year birds.
One on the estuary on Jan 6th which then roosted on the sea off Seaton;
a very white individual from Jan 18th – May 7th seen on the estuary, in the bay and off Branscombe;
a blotchy plumaged individual on the estuary on Feb 15th;
a classic but bulky bird near tram sheds on Mar 3rd;
a quite dusky-looking bird from Coronation Corner on Mar 10th;
a new bird alongside the white individual from Mar 28th – Apr 4th.

Iceland Gull *by Gavin Haig*

Glaucous Gull *Larus hyperboreus*
Rare winter visitor
As with Iceland Gull, an unprecedented year with three individuals noted, all first-year birds and all short stayers with not one seen the following day. The first was on the estuary and then Bridge Marsh on Mar 4th; then one with a hook-tipped bill on the estuary on Mar 29th, and later on Beer beach; finally, a spring bird, one over the sea on May 18th flying W at 09:45 then back E at 12:00.

Great Black-backed Gull *Larus marinus*
Fairly common winter visitor, though present throughout the year
Present all year on the estuary, beaches and sea in varying numbers though usually in double figures. This year several three figure counts, almost exclusively coinciding with periods of rough weather - highest count 350 on Mar 2nd.

Kittiwake *Rissa tridactyla*
Fairly common passage and winter visitor
All but four records concern birds flying over the sea, and no records at all in October, presumably due to calm weather conditions (also meaning no one was looking at the sea!). The biggest movement of the year occurred on Jan 1st with 151W in one hour (compare with last years peak of 580W on Nov 15th). Records away from the sea include one on Colyford Marsh on Feb 12th, one on estuary on Mar 4th-5th, with another here on May 7th. The table below shows the highest day counts in each month:

2007	Jan	Feb	Mar	Apr	May	Jun	Jul	Aug	Sep	Oct	Nov	Dec
Maxima	151	21	6	7	51	13	16	4	22		90	91

Little Tern *Sterna albifrons*
Rare passage visitor
An unprecedented two spring records of this patch rarity. Three fed off Seaton seafront all afternoon on Apr 24th, and a single bird rested and fished on the estuary off Coronation Corner on May 2nd.

Little Tern by Gavin Haig

Black Tern *Chlidonias niger*
Rare passage visitor
Three records involving three birds made for an above average year. In spring, an adult E with Common Terns on May 27th, and in autumn a juvenile lingered offshore on Aug 14th, and an adult flew W on Aug 18th.

Sandwich Tern *Sterna sandvicensis*
Fairly common passage and summer visitor, usually offshore
First record of the year: one off Seaton on Mar 28th with the final record being two here on Oct 16th (an early last date for the species). A total of 607 birds were recorded in all with peak monthly counts shown in the table below.

2007	Jan	Feb	Mar	Apr	May	Jun	Jul	Aug	Sep	Oct	Nov	Dec
Maxima			2	27	55	26	7	3	2	2		

Common Tern *Sterna hirundo*
Scarce passage visitor
First of the year: one E on Apr 24th. On May 10th a flock of 125+ flew in from the W and fed in the bay for an hour; there were another 29 individuals in May and June, with five on May 16th the spring's second highest count. In autumn, 18 birds between Jul 10th and Sep 24th – a peak of five on Aug 18th.
Uncommon away from the sea, but there were three records this year from the estuary, with singles on May 9th and Jun 6th, and two on the Colyford Common scrape, then the river, on Aug 14th.
'Commic' Tern (refers to Common or Arctic Terns seen too poorly for specific identification).
A total of 17 records involving 93 birds between May 7th and Sep 23rd with a peak of 33E on May 22nd.

Roseate Tern *Sterna dougallii*
Rare passage visitor
An excellent record for the patch occurred on May 11th at 19:50 when an adult flew slowly W past Seaton. This record coincided with a small influx of the species off Dawlish Warren the same evening, where there were up to nine individuals. This is possibly only the second record for the patch.

Arctic Tern *Sterna paradisaea*
Scarce passage visitor
Two spring records with at least two in a large flock of Common Terns off Seaton on May 10th and one E on June 5th.
In August, one W on 15th and three W on 18th. In September, after a surprise adult on the Colyford Common scrape briefly on 11th, a small passage of juveniles over the sea between 19th and 24th involved at least eight individuals, with a peak of four on 20th.

Guillemot *Uria aalge*
Fairly scarce passage and winter visitor
Small numbers regularly encountered offshore in both winter periods, with a few passage records though most passing birds are too distant to identify with certainty. No large counts this year, but four on July 1st were fairly unusual.

Razorbill *Alca torda*
Fairly scarce passage and winter visitor
Regularly seen offshore in both winter periods, with a peak of 200+ in mid-January around the Beer and Beer Head area. A total of 19 past in May, and no summer records.

Unidentified Guillemot/Razorbill *Uria aalge/Alca torda*
Most auks, especially those seen passing, are too distant to identify specifically. Largest passage occurred in late November with 596 distantly W past Seaton in three hours on 25th, with the few that did stray closer all appearing to be Razorbills. Apart from 94E on Feb 7th there were no other counts >50.

Little Auk *Alle alle*
Rare late autumn passage and winter visitor
A major displacement of this species into the North Sea took place in early November, with large numbers being recorded on the East coast. The total of seven birds seen off Seaton coincided with that event. All records in November: one W on 10th, then 2W and 1E on 12th (the latter successfully avoiding a stooping Peregrine!), 1E on 14th, 1W on 25th and finally, 1W on 30th.

Puffin *Fratercula arctica*
Rare passage and winter visitor visitor
The best year on record for this species with three sightings.
One W past Seaton on May 12th, then one on the sea seen from a fishing boat three miles off Beer on May 16th and finally, the only autumn record, one past Seaton on Sep 19th.

Rock Dove/Feral Pigeon *Columba livia*
Fairly common resident
Two populations well established at Axmouth Harbour and in the centre of Seaton. Occasional birds were picked out amongst the large flocks of migrating Woodpigeons in late autumn.

Stock Dove *Columba oenas*
Fairly scarce resident, passage and winter visitor
Breeding confirmed in Beer Quarry and on cliffs along Undercliff NNR. Largest grounded flock 50+ in Axmouth on Oct 1st.
As usual some passage occurred within the Woodpigeon autumn migration, but many must go through undetected in the more distant flocks. A total of 338 was counted passing over between Oct 29th and Nov 4th with a peak of 110 on the last date.

Woodpigeon *Columba palumbus*
Common resident breeder, passage and winter visitor
Largest grounded flock recorded 400+ in Axmouth on Oct 1st.
Autumn passage started with 538W on Oct 19th, then some impressive movements later in the month with 11,150W in three hours on 29th, 22,881W on 30th (a new local record) and 4,670W on 31st. The movement carried on into early November with 5,042W on 2nd and 5,000+W on 4th.

Collared Dove *Streptopelia decaocto*
Fairly common resident breeder
Regularly found in inhabited areas but no counts of more than ten made.

Turtle Dove *Streptopelia turtur*
Rare/scarce passage visitor
After four records last year, only one in 2007 with a single on telephone wires by Beer Cemetery on the unusual date of Jul 16th.

Cuckoo *Cuculus canorus*
Rare/scarce passage visitor, has bred
By recent standards a good year with six records involving probably five birds, all in spring. The first flew E over Beer Head on Apr 19th, then a single calling near Branscombe Mouth on Apr 24th, with presumably the same bird heard here again on 27th. The final three records were also heard-only, with singles in Seaton on May 19th, Colyton on May 23rd and Combpyne on one morning in early June.

Barn Owl *Tyto alba*
Scarce resident breeder
One pair bred successfully in Beer, raising four young.
In both first and second winter periods most often encountered along A3052 west of Tower Services, and around Lower Bruckland Ponds; also the odd record from Musbury. Still rare in river valley S of A3052, with only a single record here – one hunting over Black Hole and Axe Marshes on Nov 23rd.

Little Owl *Athene noctua*
Scarce resident breeder
Breeding not confirmed this year despite two pairs and a single bird present in the Musbury area throughout the summer months, all staying loyal to their sites. Also recorded from a couple of sites in Colyton but more sporadically.

Little Owl *by Steve Waite*

Tawny Owl *Strix aluco*
Fairly common resident breeder
Heard regularly in Seaton and throughout the area in suitable habitat. Although it probably breeds at many localities, only confirmed in Colyton.

Short-eared Owl *Asio flammeus*
Rare passage visitor
Recorded for the second year in a row, begging the question of whether increased observer coverage is proving that they are not as scarce as first thought?
One flew high north over Colyford Common on Oct 5th, and two birds were seen on and off over Colyford Marsh and adjacent fields from Nov 9th – 13th, with (different?) singles also observed here on Nov 23rd and Dec 18th. On the evening of Nov 12th one of the Colyford birds flew high out to sea over Seaton Hole, but after about a mile U-turned back to the marshes.

Swift *Apus apus*
Fairly common summer and passage visitor; breeds
Earliest records were of one over Colyford Common on Apr 23rd and six arriving off the sea at Seaton on May 1st; largest count was c80+ hawking flying ants over Seaton in mid-late July and latest were singles at Beer Head Aug 22nd and Colyford Common on Aug 31st. Breeding appears to have been disappointing: although 25-30 were seen on several occasions around Colyton Parish church only one 'screaming' family flight was reported.

Kingfisher *Alcedo atthis*
Fairly scarce breeding resident and probable winter visitor
Regularly seen in ones or twos on the estuary, marshes and higher reaches of the Coly but up to four in the harbour area and Seaton Marshes in February and November, four along the Coly on Apr 5th and five ringed at Colyford Common LNR on July 10th, the latter showing that sightings of singles at the reserve may well involve several individuals.

Hoopoe *Upupa epops*
Rare vagrant from Europe
A single spring record of one in a private garden in Beer on Apr 26th which, as usual, managed to avoid the local birding community!

Hoopoe *by Andrew Coates*

Green Woodpecker *Picus viridis*
Fairly scarce resident breeder
Possibly under-recorded: occasional reports from the reserves and heard regularly in Colyton, particularly near the river. One private Colyton garden was visited regularly during the breeding season by both male and female for ants and one juvenile was seen there on Jul 9th.

Great Spotted Woodpecker *Dendrocopus major*
Fairly scarce resident breeder and probable passage visitor
Occurs throughout the area in suitable habitat. Breeding records: three nests with young at Holyford Woods on Apr 29th and three young seen leaving a nest at Heathayne along the Coly on May 29th. Possible migrants: singles on Beer Head Aug 16th and Sep 12th and at Haven Cliff on Oct 30th.

Skylark *Alauda arvensis*
Fairly common passage and winter visitor; present in summer and probably breeds
Breeding: heard regularly in ones or twos at Colyford Marsh but a count of ten there in late August probably reflects early passage rather than breeding success.
Winter: the only significant winter record is of 47 in the Haven Cliff area on Jan 13th.
Passage: noted from early October to mid November with a peak count of 378W plus 50 grounded birds at Haven Cliff on Oct 30th and 450+ on farmland at Axmouth on Oct 31st.

Sand Martin *Riparia riparia*
Fairly common passage and summer visitor; breeds
Earliest record was of one at Seaton Marshes on Mar 10th. Breeding: both colonies in the recording area (nr. Ratshole Gate on the Coly and just south of the A3052 on the same river) were active although numbers were not estimated.
Largest counts were of 120 in the Axe estuary on Apr 18th and c. 100 at Colyford Common during the last week of August.

Swallow *Hirundo rustica*
Common passage and summer visitor; breeds
Earliest record was of one over Borrow Pit and Seaton Marshes on Mar 19th; latest was one at Beer Head on Oct 29th. Spring passage was noted especially on Apr 25th (c180/hrW) and between May 7th and 10th (described as 'heavy') and again on 19th. Maximum count in the late summer/autumn was 100+ at Beer Head on Jul 28th.

Swallow by Roger Boswell

House Martin *Delichon urbica*
Common summer and passage visitor; breeds
The first spring record was one at Axmouth on Mar 25th but breeding birds in Colyton arrived later this year (Apr 23/24th) and flocks of 40, 34 and 23 were seen arriving off the sea during a two hour period early morning as late as May 28th. Notable autumn totals were 250+ at Haven Cliff on Sep 18th, 400+ at Beer Head on Sep 28th and 500+ for much of the day over the River Axe on Oct 1st. Latest date: one flying south over Seaton on Nov 8th.

Tree Pipit *Anthus trivialis*
Fairly scarce passage visitor
Only one spring record (cf. seven in 2006) with a single N over Beer Head on Apr 7th. A better autumn with 23 birds recorded between Aug 8th and Oct 3rd. Most records concern overflying birds from Beer Head or Haven Cliff, with a maximum of six at the former on Sep 9th. Records from river valleys include one W over Colyford Common on Sep 27th and the year's final record of one over Axe Marsh on Oct 3rd.

Meadow Pipit
Anthus pratensis
Common passage and winter visitor; breeds
Bred on Colyford Marsh with young seen after at least four singing males were present. No large wintering flocks noted though Colyford Common and Haven Cliff seem to be the favoured sites.

First spring migrant: one W on Feb 24th; there were two dates with especially large spring passages but counting always near impossible as the passage occurs on such a broad front, estimates of 400+ on Mar 26th and several thousands over on Apr 2nd are probably nowhere near the actual figure!

Meadow Pipit *by Roger Boswell*

First autumn migrants seen on Beer Head, with 15 here on Sep 5th, then lots of overhead passage throughout rest of September and October with counts >100 on 13 days, the biggest movement occurring on Sep 22nd when large flocks constantly streamed W over Seaton, again easily several thousand birds were involved but counting was near impossible!

Rock Pipit *Anthus petrosus*
Fairly common passage and winter visitor; breeds
Four pairs found on beach between Beer Head and Branscombe during a cliff nesting survey in April.

Regularly present during the winter around the estuary, particularly at Colyford Common with up to 15 present, though most of these showed features associated with *A. p. littoralis* (see below). Other notable winter concentrations at Axmouth Harbour, Seaton Hole and Beer Beach.

A. p. littoralis (**Scandinavian Rock Pipit**)
Scarce winter visitor?
Although it has been thought that most/all the wintering Rock Pipits at Colyford Common were of the race *littoralis*, this is the first year that several of the birds moulted into summer plumage, confirming this. Up to four were seen in this plumage state from mid February onwards, the last record being of two in breeding plumage on Apr 12th. Also, a Rock Pipit trapped and ringed here on Feb 1st showed features associated with this race.
Records of A. p. littoralis pending consideration by DBRC, except the trapped bird, which has been accepted.

Water Pipit *Anthus spinoletta*
Scarce passage and winter visitor
The Axe estuary is one of the best localities in Devon to see this species, though for the second successive year no double figure counts were made (cf. 22 in 2005).
Only recorded from Colyford Marsh/Common as usual. Highest first winter count on Feb 16th with six, otherwise counts of three birds regular; last spring sighting was of two on Mar 22nd. First returning bird seen on Oct 19th, with up to three seen until year end with a peak of five on Dec 27th.
Three were trapped and ringed by AERG at Colyford Common during the second winter period.

Water Pipits *by Fraser Rush*

Yellow Wagtail *Motacilla flava flavissima*
Scarce spring and fairly scarce autumn passage migrant
Only four spring sightings, all over Beer Head: two on Apr 23rd with singles on May 24th and 25th.
First autumn record was 14 at Beer Head on Aug 9th, then regularly seen mostly here or in the river valley until late September with highest counts of 39 on Aug 24th and 32 on Sept 5th (cf. 2006 peaks of 50 – twice – and 60). Last record of year: five at Beer Head on Oct 3rd.

M. f. flava (Blue-headed Wagtail)
Rare passage visitor
A smart male at Colyford Marsh briefly on evening of May 5th.

Yellow Wagtail by Roger Boswell

Grey Wagtail *Motacilla cinerea*
Fairly scarce resident breeder and passage migrant
Breeding confirmed on Coly and Axe. In winter often seen in suitable areas though no high counts were made.
A light autumn passage was noted, with a very low peak count of five W on Sept 17th.

Pied Wagtail *Motacilla alba yarrellii*
Common resident breeder, passage and winter visitor
Young were seen at a few sites in late summer, including both Colyford and Seaton Marshes. Common throughout the area in both winter periods, the largest congregations occurring pre-roost, mainly in Seaton and around the estuary, with counts of 60+ on Colyford Marsh in both winter periods.
During autumn visible migration counts, birds flying over are recorded as Pied/White Wagtails but as very little counting was conducted this year no large counts were made, with 30 the sum total of the entire autumn passage (cf. 2006, when 226 flew W on one day, with counts >100 on three other dates).

M. a. alba (White Wagtail)
Scarce passage visitor
Up to two on Colyford Marsh from Feb 21st to Mar 15th were probably over-wintering birds as opposed to early spring migrants; two at Colyton WTW on Mar 13th may have been different individuals. A further six spring birds were noted, all in April. Only autumn record: two at Beer Head on Oct 5th.

Dipper
Cinclus cinclus
Rare/scarce visitor, though probably breeds nearby
Three records this year from the Coly and Stafford Brook. At Umborne Bridge on Jan 29th, Stafford Brook on Feb 28th (both singles) and at Heathhayne between Nov 1st and 6th, when two were seen.

Dipper *by Karen Woolley*

Wren *Troglodytes troglodytes*
Common resident breeder
Very common in almost all habitats throughout the area.

Dunnock *Prunella modularis*
Common resident breeder
Not so common in woodland as in town and country gardens.

Robin *Erithacus rubecula*
Common resident breeder
One of few species to appear in consistent numbers in all habitats in early winter BTO Atlas surveys.

Black Redstart
Phoenicurus ochruros
Scarce passage and winter visitor
Seven records from Beer and Seaton in the first winter period between Jan 1st and Mar 7th. Over 20 records from Oct 17th with Haven Cliff, Axmouth Harbour and Durley Road as additional sites. At least four wintering in December, with a male and female on Beer beach plus two females at Axe Yacht Club on 31st.

Black Redstart *by Steve Waite*

Redstart *Phoenicurus phoenicurus*
Fairly scarce passage visitor

Eight fewer birds than last year but a probable 27 is still significant in the Devon context. 20 in spring between Apr 7th and May 16th included a peak of four at Beer Head on Apr 15th. Most records were from here but also from Haven Cliff and, oddly, Boshill Cross where a female was on the road on Apr 24th. All seven autumn birds were at Beer Head between Aug 17th and Sept 30th, with two on Aug 24th.

Whinchat *Saxicola rubetra*
Fairly scarce passage visitor

Very similar numbers to 2006. Four spring birds: Seaton Marshes on Apr 17th, the estuary saltmarsh on Apr 25th, Beer Head on May 6th and Seaton Marshes again on May 13th. A sprinkling through the autumn, mostly at Beer Head or the estuary marshes, though the last was at Haven Cliff on Oct 8th.

2007	Jan	Feb	Mar	Apr	May	Jun	Jul	Aug	Sep	Oct	Nov	Dec
Maxima				1	1			2	1	1		
Bird-days				2	2			20	6	4		

Stonechat
Saxicola torquata
Fairly common passage and winter visitor; breeds

Breeding season records: two pairs at Black Hole Marsh on May 15th and three recently fledged juveniles there on Jun 6th; Beer Head, a family party on Aug 8th; also three sites in the Undercliffs at Bindon, the Plateau and Rousdon.

It may appear almost anywhere in winter but autumn passage was most often recorded from Beer Head with many between Aug 9th and Oct 18th including 19 on Oct 4th.

Stonechat *by Karen Woolley*

Wheatear *Oenanthe oenanthe*
Fairly common passage visitor
The first, an early bird at Beer Head on Mar 8th, was nearly two weeks ahead of the next! Spring passage ended on May 22nd with peak individual counts of 18 at Beer Head on Apr 16th and 14 at Seaton Marshes on Apr 2nd. A few Greenland race birds were recorded in May.
Autumn passage began with one at Beer Head on Jul 18th, though numbers were lower than last year with highest counts at Beer Head being 34 on Aug 21st and 19 on Sep 24th. The last of the year were three at Haven Cliff on Oct 18th.

2007	Jan	Feb	Mar	Apr	May	Jun	Jul	Aug	Sep	Oct	Nov	Dec
Maxima			7	26	3		2	40	19	7		

Wheatear *by Roger Boswell*

Ring Ouzel *Turdus torquatus*
Scarce passage visitor
Fewer records than last year with only one in spring: at Beer Head on Apr 15th. Recorded on six dates in autumn between Sep 29th and Nov 2nd, with a peak of three on Oct 26th, all at Beer Head.

Blackbird *Turdus merula*
A common resident breeder, passage and winter visitor
Consistently conspicuous in all habitats during early winter Atlas counts at a time when up to 25 fed in a small Combpyne orchard.

Fieldfare *Turdus pilaris*
Fairly common winter visitor
Early in the year the highest counts were 150 at Bindon Barns on Jan 13th and 350+ south of Whitford on Jan 25th. Last recorded on Mar 30th near Holyford Woods. In the second winter period the first birds were seen on Oct 2nd and despite numerous records the highest count was only 100, at Colyford Common on Nov 8th.

Song Thrush *Turdus philomelos*
Common resident breeder, passage and winter visitor
In Combpyne and the Undercliffs four can sometimes be heard singing at the same time. Between Dowlands and Rousdon on Apr 13th there were more singing Song Thrushes (13) than Blackbirds.

Redwing
Turdus iliacus
A fairly common winter and passage visitor
Three treble figure counts in the period up to Mar 30th with more than 400 south of Whitford on Jan 25th. First autumn record was at Colyford Common on Oct 3rd, after which migrants were often heard overhead at night and visible migrants seen over Haven Cliff, though no notable counts.

Redwing by Fraser Rush

Mistle Thrush *Turdus viscivorus*
Fairly scarce passage and winter visitor; breeds
Breeding season records from Axmouth, Combpyne, Holyford Woods, Horriford Farm and the Undercliffs with a definite breeding record of a nest near the river Coly on Apr 21st. Four apparent migrants flew W together at Haven Cliff on Nov 2nd.

Cetti's Warbler *Cettia cetti*
Scarce resident breeder
Early song on Feb 2nd at Seaton Marshes and Feb 4th at Colyford Common. Perhaps five breeding pairs to the west of the Axe with two at Colyford, two at Seaton and one in between. Three recorded at Colyford Common on Oct 1st. Six different individuals were trapped and ringed through the year.

Grasshopper Warbler *Locustella naevia*
Scarce passage visitor
An even better flurry of spring birds than last year with eight between Apr 15th and 29th, mainly from Beer Head but also from Musbury and Beer Quarry, where two were 'reeling' on Apr 22nd. Two autumn records as in 2006, from Colyford Common on Aug 13th and Beer Head on Aug 26th.

Sedge Warbler *Acrocephalus schoenobaenus*
Fairly common passage and summer visitor; breeds
First record from the Axe on Apr 16th. Proof of breeding, with a family at Lower Bruckland Ponds on Jul 11th, but still under-recorded. Three autumn records from Haven Cliff and Beer Head between Aug 6th and Sep 8th with four at the latter on Aug 11th.

Reed Warbler *Acrocephalus scirpaceus*
Fairly common passage and summer visitor; breeds
Despite good populations singing in estuary reed beds, particularly adjacent to the observation platform at Colyford Common, no counts were reported. The only other noted breeding indication was of song at Lower Bruckland Ponds on Jul 11th. Present between Apr 21st and Oct 6th.

Blackcap *Sylvia atricapilla*
Common passage and summer visitor, scarce winter visitor; breeds
A sign of its abundance in the Undercliffs was that 27 were heard between Bindon and Dowlands on Apr 13th. Some indication of its winter status was a record of seven in a Seaton garden on Dec 11th. A good autumn count of 35+ was made at Haven Cliff on Sep 10th, while the peak at Beer Head was 17 on Sep 12th.

Garden Warbler *Sylvia borin*
Fairly scarce summer and passage visitor; probably breeds
First record was of song at Colyford Common on Apr 26th and a sign of possible breeding was a singing male at Heathhayne Farm, by the Coly, on May 23rd. Six autumn records, less than last year, between Aug 28th and Sep 12th at Beer Head with birds also at Beer Cemetery on Aug 25th and at Haven Cliff four days later.

Lesser Whitethroat *Sylvia curruca*
Fairly scarce passage and summer visitor; breeds
The first of the year at Colyford Common on Apr 16th, 16 days earlier than last year, was followed by others at Axmouth, Beer Head, Beer Quarry and Whitford Bridge, with a rare proof of breeding when a pair were seen feeding young on the western slopes of Beer Head on Jun 5th. Possible breeding again at Haven Cliff. Autumn records from Beer Head on six days between Aug 8th and Sep 9th, all singles apart from two on the latter date.

Whitethroat *Sylvia communis*
Fairly common passage and summer visitor; breeds
An early first from Seaton Marshes on Mar 31st. This is one of many species where BTO Atlas work should give a better indication of breeding numbers but as in previous years Stepps Lane and Haven Cliff were good sites. Some 30 autumn records from Beer Head between Aug 8th and Oct 3rd with a peak of five on Aug 26th.

Dartford Warbler *Sylvia undata*
Rare passage visitor

Records again from unexpected sites with one in weak song at Colyford Common on Mar 1st, a juvenile at Haven Cliff on Jul 30th and then a run of records from Beer Head on seven dates from Aug 11th, when 3 juveniles were found frequenting a barley crop; apart from an adult on Aug 16th all others were single juveniles in the same area, with the last on Oct 5th. The suspicion is that the Beer Head records must have stemmed from some not-too-distant breeders!

Dartford Warbler by Karen Woolley

Pallas's Warbler *Phylloscopus proregulus*
Rare vagrant

The first confirmed record of this species on patch was on the afternoon of Oct 22nd when one showed briefly around Axmouth churchyard. It was seen briefly again the following morning.
Accepted by DBRC

Yellow-browed Warbler *Phylloscopus inornatus*
Rare vagrant

An amazing three records this year brings the area total to five. As the 'Axe Estuary Birds' newsletter no. 71 commented: 'it's a treat to be able to gloat about having two species of small, stripey, *Phylloscopus* warblers on our year list.' There were three records, all in October: at Beer Head on 3rd, at Lower Bruckland Ponds from 9th to 11th and at Cottshayne Hill on 30th. The latter bird was found during a woodland bird survey, when it was in willows with Blue Tits!
All records accepted by DBRC

Chiffchaff *Phylloscopus collybita*
Common passage and summer visitor; fairly scarce winter visitor; breeds

The peak first winter count was 15 at Colyton WTW on Jan 21st and the first song was heard at Axe Marsh on Mar 6th. A fledgling in Bovey Lane on May 20th provided proof of breeding. The success of this species was indicated by 22 singing in the Undercliff between Dowlands and Rousdon on Apr 13th. Autumn passage at Beer Head noted between Aug 5th and Oct 5th with peaks of 15 on Aug 20th, 21 on Sep 15th and 30+ on Aug 4th. Also in early October up to ten were at Lower Bruckland Ponds. The last song of the year was on Oct 11th at Whitford Bridge. No counts were recorded in the second winter period though birds were, as usual, present in suitable habitat.

Iberian Chiffchaff *Phylloscopus ibericus*
Rare vagrant from S Europe
At Beer Head an elusive male found at 11:00 on Apr 28th spent the rest of the day singing from the undercliff blackthorn, but was gone the following morning. A new species for the recording area, and only the fourth in Devon.
Accepted by BBRC

Willow Warbler *Phylloscopus trochilus*
Common passage and summer visitor; breeds
The first was one in Beer on Mar 25th. Spring passage was noted at Beer Head between Apr 13th and May 5th with a peak of 35 on the afternoon of Apr 13th. Peculiar record from Cottshayne woods on May 31st when normal song was followed by a Chiffchaff version. Probably breeding for the first time in Colyton Community Woodland. Autumn passage records from Beer Head between Aug 8th and Sep 8th peaked at 27 on Aug 13th.

Goldcrest *Regulus regulus*
Fairly common resident breeder, passage and winter visitor
Preliminary winter work for the BTO winter Atlas shows how widespread this species is with records from every 2 x 2km tetrad and as many birds in deciduous as coniferous woods. On passage it occurs along the coast at Haven Cliff and Seaton Hole as well as at Beer Head where it is more frequent in the autumn.

Firecrest *Regulus ignicapillus*
Scarce passage and winter visitor, possibly breeds
In the first winter period there were three records from Jubilee Gardens at Beer where two were present on Mar 30th. Two were back at this site on Nov 13th, after autumn records as follows: Beer Head on Oct 4th and 26th, and Nov 7th, Axmouth on Oct 23rd and Haven Cliff on Oct 29th. A sighting in Cottshayne Hill Woods on Jun 30th suggests the possibility of breeding.

Firecrest by Karen Woolley

Spotted Flycatcher *Muscicapa striata*
Fairly scarce passage and summer visitor; breeds
A single bird at Beer Head on Apr 28th was an early spring record. Others were noted there in mid to late May in small numbers. Breeding was reported from the banks of the River Coly near Heathhayne and juveniles were noted there, in Morganhayes Covert and in Beer Cemetery. The largest autumn count was ten on the eastern slopes of Branscombe on Aug 26th and the last records were of singles at Beer Head and Colyford Common on Sep 15th.

Pied Flycatcher *by Steve Waite*

Spotted Flycatcher *by Richard Halliwell*

Pied Flycatcher *Ficedula hypoleuca*
Scarce passage visitor
The only spring record was one at Beer Head on Apr 13th. An excellent autumn passage, with three at Beer cemetery fields on Aug 24/25th and thereafter singles at Beer Head on Aug 25/26th and Sep 9th, plus one at Colyford Common on Sep 15th (a first for the reserve).

Bearded Tit *Panurus biarmicus*
Rare passage visitor
A record of three, including an adult male, roosting in reeds at Colyford Marsh on Oct 12th was a first for the reserve, and the first locally in recent times. There was no sign of them at dawn the next day.

Long-tailed Tit *Aegithalos caudatus*
Fairly common resident breeder
Counts of ten or more were noted at Seaton Marshes in mid June (10) and late November (12), Beer Head Oct 12th (10) and 18th (16), and Colyford Common in mid September and mid October (12), with a peak of 27 in the last week of October.

Blue Tit *Cyanistes caeruleus*
Common resident breeder
Widespread and commonly encountered particularly at bird feeders in gardens. Evidence from nest box studies in Holyford Woods suggests that late nesting birds were less successful than earlier nesters, presumably as a result of the cold wet weather during the summer.

Great Tit *Parus major*
Common resident breeder
Distribution very much as Blue Tit. It appears to have been a succesful breeding year.

Coal Tit *Periparus ater*
Fairly common resident breeder and scarce passage visitor
Present throughout the area in small numbers; four at Seaton Marshes in early July.

Marsh Tit *Poecile palustris*
Fairly scarce resident breeder
Rarely noted away from favoured woodland habitats, where it can be quite numerous. Rarely encountered away from favoured woodland but one was seen at Seaton Marshes in April. Breeding: young were seen near Heathayne Farm on the banks of the River Coly on May 29th; a juvenile was seen in a private garden in Colyton on Jul 1st.

Treecreeper *Certhia familiaris*
Fairly scarce resident breeder
Breeding was reported from the Heathayne area of the River Coly where young were seen on May 25th and up to three birds were at the same location in October and November. A party of four was seen in Morganhayes Covert on Oct 31st.

Jay *Garrulus glandarius*
Fairly scarce resident; probably breeds
Under-recorded; the only records are of one at Seaton Marshes in February and singles at Colyford Common during the last week of September and first week of October.

Magpie *Pica pica*
Common resident breeder
Under-recorded; the only 'gathering' noted was of 12 at Seaton Marshes towards the end of February. In Colyton one was seen with a slow-worm on Apr 13th (the slow worm was 'rescued' by the householder and released later).

Jackdaw *Corvus monedula*
Common resident breeder
Breeding: always very much in evidence at Seaton Hole. 60 occupied nests were found on cliffs between Seaton and Branscombe during a cliff nesting survey in April. Autumn movements included an exceptional 1200+W over Seaton on Oct 18th, 1000+ over Beer Head on Oct 21st and 800W over Seaton on the same day.

Rook *Corvus frugilegus*
Common resident breeder
There are several small rookeries within the area but this species is probably under-recorded, particularly when forming mixed flocks with Carrion Crows. The largest counts were of 80 at Colyford Marsh during the last week of September and 70 at Seaton Marshes at the beginning and end of the year.

Carrion Crow *Corvus corone*
Common resident breeder
Breeds throughout the area and is very much in evidence in the piggeries, and in low lying fields around the estuary in flocks which are often in excess of 100 birds.

Raven *Corvus corax*
Fairly scarce resident breeder
Seen regularly throughout the year along the coast and inland. Two nest sites were found between Seaton Hole and Branscombe. A party of seven was seen flying W low over Colyton on Jan 12th and six near Colyton Picnic Site on Feb 7th; 15 were reported near the piggeries close to Hangman's Stone on Feb 5th.

Raven *by Bob Hastie*

Starling *Sturnus vulgaris*
Common resident breeder, passage and winter visitor
Widely but thinly spread within the area during the breeding season. Post-breeding flocks of 100-150 during June to August and winter flocks of c. 200-250 were reported from Colyford Common. From mid October until late November Starlings roosted in the estuary reed beds, with numbers peaking at c. 5000.

House Sparrow *Passer domesticus*
Fairly common resident breeder
Abundant throughout the area all year, with breeding confirmed in several rural and urban locations.

Chaffinch *Fringilla coelebs*
Common resident breeder, passage and winter visitor
Seen throughout the year. Biggest wintering flocks recorded east of Axmouth and on Allhallows estate with 300 counted here on Feb 5th; in second winter period also on seed at Colyford Common with 250+ here in Nov/early Dec.
Light spring movement on Mar 27th with 56E over Beer Head. Large movements took place in late autumn, with a huge count of 2,307W in four hours on Oct 30th. Other large counts include 750+W on Oct 29th, 558W on Oct 31st, and in November 295W on 3rd and 106W on 6th. It should be noted that many days with large passages went by uncounted.

Brambling *Fringilla montifringilla*
Scarce passage and winter visitor
Only records for first winter period were from a Seaton garden, with singles on Jan 7th, 9th and 21st.
First record of second winter period was one during a visible migration watch on Oct 29th, then a further 31 counted during similar watches through October and November, peaking at ten on Nov 3rd.
In winter, one to three at Colyford Common in late November, up to ten at Rousdon drying barns through December, with occasional singles elsewhere.

Greenfinch *Carduelis chloris*
Common resident breeder, passage and winter visitor
Still appears to be scarcer than usual in gardens in both winter periods with no large flocks reported.
Seen in very small numbers during autumn migration watches, with a high of 89W on Oct 30th.

Goldfinch *Carduelis carduelis*
Common resident breeder, passage and winter visitor
Young seen at several sites, and the highest winter count was 70+ in a Seaton garden on Jan 9th.
A light spring passage with a low peak of 27W over Beer Head on Apr 25th; autumn passage went over largely uncounted, with the highest count made being 46W on Oct 30th.

Siskin *Carduelis spinus*
Fairly scarce passage and winter visitor, scarce in some years; rare breeder
Breeding confirmed for the first time. After an unseasonal male in a private garden in Colyton on June 26th, two juveniles were photographed in Morganhayes Woods on Aug 18th. Very scarce in first winter period with only two records of five birds. Much more evident in second winter period with a flock of 30+ at Heathayne Bridge on the Coly during November/December and large numbers in Morganhayes Woods and at Bovey Down.

Siskin by *Karen Woolley*

Overhead passage comprised 158 counted between Sep 17th and Nov 23rd, the largest flock being 28W over Haven Cliff on Nov 21st.

Serin *Serinus serinus*
Rare passage visitor/vagrant from Europe
One flew low E along Haven Cliff on Nov 6th at 07:30 calling, this being only the second confirmed record of this species in the area.
Accepted by DBRC

Linnet *Carduelis cannabina*
Common resident breeder, passage and winter visitor
Breeding confirmed with young seen at Colyford Common, Beer Head and Haven Cliff, and probably many more sites!
Largest grounded flock at Haven Cliffs in October with 1000+ present during middle of month. Largest wintering flocks of 200 near Pratt's Hill, Colyton on Feb 27th and Haven Cliff on Nov 29th.
Overhead migrants only counted on two days in autumn with 167W on Oct 30th and 103W on Nov 2nd.

Lesser Redpoll *Carduelis cabaret*
Scarce passage and winter visitor
Surprisingly not recorded until Oct 1st with one in Cottshayne Hill Woods, with eight here again on Oct 30th and one on Nov 29th.
Also seen in autumn during visible migration watches: a total of 38 over (most W), with a peak of 12 on both Oct 29th and Nov 23rd. Many many more would have passed through unseen though!

Bullfinch *Pyrrhula pyrrhula*
Fairly scarce resident breeder and passage visitor
Present in woods, parks and gardens throughout the patch in small numbers all year, with young seen at several sites from mid summer.

Lapland Bunting *Calcarius lapponicus*
Rare passage visitor
One heard flying W over Haven Cliff at 07:10 on Oct 30th constitutes the first record for the area.
Accepted by DBRC

Yellowhammer *Emberiza citrinella*
Fairly scarce passage and winter visitor, present throughout the year; breeds
Breeding confirmed on many farmland sites, with young seen from mid summer.
Largest flocks in both winter periods occurred on farmland around Axmouth/Rousdon area with 50 here on Jan 13th; also on farmland north and west of Beer.

Reed Bunting
Emberiza schoeniclus
Fairly common resident breeder, fairly scarce passage and winter visitor
14-16 singing males in the river valley south of A3052, with several successful breeding pairs noted.
Wintering birds recorded from farmland E of Axmouth and near Beer.
Definite migrants were recorded only from Haven Cliff with a total of 18 birds between Oct 29th and Nov 11th with peak of eight W on Oct 30th.

Reed Bunting *by Roger Boswell*

The Axe Estuary and Seaton Bay Bird Report 2007

Axe Estuary and Seaton Bay Birds – Species List

All BOU Category A, B and C taxa that are known to have occurred in the recording area are listed here, and those occurring in the current year (2007) are shown in bold. For all others, the year and month when they were last recorded (if known) is given. Where taxonomically distinct forms (either nominate race or subspecies) have occurred, they are indented. Devon rarities are marked with an asterisk (*). National (BBRC) rarities are marked with a double asterisk (**).
In both cases, full supporting notes should accompany any records and be sent to the Devon Bird Recorder, whose name and address can be found on the Contacts page.

Mute Swan
Bewick's Swan   Oct 2001
Whooper Swan
Bean Goose   *
   'Tundra' Bean Goose   * Jan 2005
   Pink-footed Goose   Jan 1963
White-fronted Goose   *
   'European' White-fronted Goose   *
Greylag Goose
Canada Goose
Barnacle Goose
Brent Goose
   'Dark-bellied' Brent Goose
   'Pale-bellied' Brent Goose
Egyptian Goose
Ruddy Shelduck   Jun 2006
Shelduck
Mandarin
Wigeon
Gadwall
Teal
Mallard
Pintail
Garganey
Shoveler
Pochard
Ferruginous Duck   *   Apr 1968

Tufted Duck
Scaup   Apr 1994
Eider
Long-tailed Duck
Common Scoter
Surf Scoter   *
Velvet Scoter
Goldeneye
Smew   Jan 2004
Red-breasted Merganser
Goosander
Ruddy Duck   May 2003
Red-legged Partridge
Grey Partridge
Quail
Pheasant
Red-throated Diver
Black-throated Diver
Great Northern Diver
Little Grebe
Great Crested Grebe
Red-necked Grebe   Nov 2006
Slavonian Grebe
Black-necked Grebe
Fulmar
Sooty Shearwater
Manx Shearwater

77

Balearic Shearwater
Storm Petrel
Leach's Petrel * Dec 2006
Gannet
Cormorant
Shag
Bittern Dec 1960
Little Bittern ** May 1869
Night Heron * Apr 2006
Squacco Heron ** May 1997
Cattle Egret **
Little Egret
Great White Egret * Sep 2005
Grey Heron
Purple Heron * Mar 2006
Black Stork **
White Stork * Mar 2006
Glossy Ibis ** Dec 1964
Spoonbill
Honey Buzzard
Red Kite
Marsh Harrier
Hen Harrier
Goshawk *
Sparrowhawk
Buzzard
Osprey
Kestrel
Merlin
Hobby
Gyr Falcon ** Jun 1882
Peregrine
Water Rail
Spotted Crake * Aug 1947
Corncrake
Moorhen
Coot
Common Crane * Dec 2004
Oystercatcher

Avocet Dec 2005
Stone-curlew
Little Ringed Plover
Ringed Plover
Dotterel Sep 2005
Golden Plover
Grey Plover
Lapwing
Knot
Sanderling
Western Sandpiper ** Sep 1973
Little Stint
Temminck's Stint *
Pectoral Sandpiper *
Curlew Sandpiper
Dunlin
Buff-breasted Sandpiper * Oct 2006
Ruff
Jack Snipe
Common Snipe
Woodcock
Black-tailed Godwit
Bar-tailed Godwit
Whimbrel
Curlew
Common Sandpiper
Green Sandpiper
Spotted Redshank
Greenshank
Wood Sandpiper
Redshank
Turnstone
Wilson's Phalarope ** Oct 1991
Grey Phalarope Nov 2005
Pomarine Skua
Arctic Skua
Long-tailed Skua * Aug 2006
Great Skua
Mediterranean Gull

Laughing Gull **
Little Gull
Sabine's Gull Nov 2005
Bonaparte's Gull **
Black-headed Gull
Audouin's Gull **
Ring-billed Gull *
Common Gull
Caspian Gull
Lesser Black-backed Gull
   'Scandinavian' LBBG (*L. f. intermedius*) *
Yellow-legged Gull *
Herring Gull
Iceland Gull
Glaucous Gull
Great Black-backed Gull
Kittiwake
Little Tern
Gull-billed Tern ** May 2006
Caspian Tern ** Jul 1966
Black Tern
Sandwich Tern
Common Tern
Roseate Tern *
Arctic Tern
Guillemot
Razorbill
Black Guillemot * Mar 2005
Little Auk
Puffin
Feral Rock Dove
Stock Dove
Woodpigeon
Collared Dove
Turtle Dove
Cuckoo
Barn Owl
Little Owl

Tawny Owl
Short-eared Owl
Nightjar 2005
Chimney Swift ** Oct 1999
Swift
Alpine Swift * Apr 2006
Kingfisher
Bee-eater * Jun 1949
Hoopoe
Wryneck Oct 2006
Green Woodpecker
Great Spotted Woodpecker
Lesser Spotted Woodpecker Oct 2006
Woodlark Jun 1947
Skylark
Sand Martin
Swallow
House Martin
Richard's Pipit * Nov 2005
Tree Pipit
Meadow Pipit
Rock Pipit
   'Scandinavian' Rock Pipit (*A. p. littoralis*)*
Water Pipit
Yellow Wagtail
   Blue-headed Wagtail
Grey Wagtail
Pied Wagtail
   White Wagtail
Dipper
Wren
Dunnock
Robin
Nightingale
Black Redstart
Redstart
Whinchat
Stonechat

Wheatear
Black-eared Wheatear ** May 1947
Ring Ouzel
Blackbird
Fieldfare
Song Thrush
Redwing
Mistle Thrush
Cetti's Warbler
Grasshopper Warbler
Sedge Warbler
Reed Warbler
Great Reed Warbler ** May 1992
Blackcap
Garden Warbler
Lesser Whitethroat
Whitethroat
Dartford Warbler
Pallas's Warbler *
Yellow-browed Warbler *
Hume's Warbler ** Dec 2005
Wood Warbler Aug 2006
Chiffchaff
 P. c. abietinus/tristis * Jan 2005
Iberian Chiffchaff **
Willow Warbler
Goldcrest
Firecrest
Spotted Flycatcher
Pied Flycatcher
Bearded Tit
Long-tailed Tit
Blue Tit
Great Tit
Coal Tit
 'Continental' Coal Tit * Nov 2005
Marsh Tit
Nuthatch
Treecreeper

Red-backed Shrike
Woodchat Shrike * Jun 1958
Jay
Magpie
Nutcracker ** Sep 1968
Jackdaw
Rook
Carrion Crow
Raven
Starling
Rose-coloured Starling * Jun 2000
House Sparrow
Tree Sparrow Dec 1976
Chaffinch
Brambling
Serin *
Greenfinch
Goldfinch
Siskin
Linnet
Twite *
Lesser Redpoll
Crossbill Sep 2006
Bullfinch
Hawfinch Apr 2006
Lapland Bunting *
Snow Bunting Mar 2006
Yellowhammer
Cirl Bunting 1964
Ortolan Bunting * Aug 2006
Reed Bunting
Corn Bunting * Apr 1989

The Axe Estuary and Seaton Bay Bird Report 2007

Maps

The following pages show maps for the area covered in this report.

The first gives the geographical extent of the recording area, with annotations for specific locations often referred to within the report.

The second map is a more detailed view of the Axe Estuary, the East Devon District Council Local Nature Reserves and other significant birding locations around the estuary.

Axe Estuary and Seaton Bay Bird Report 2008

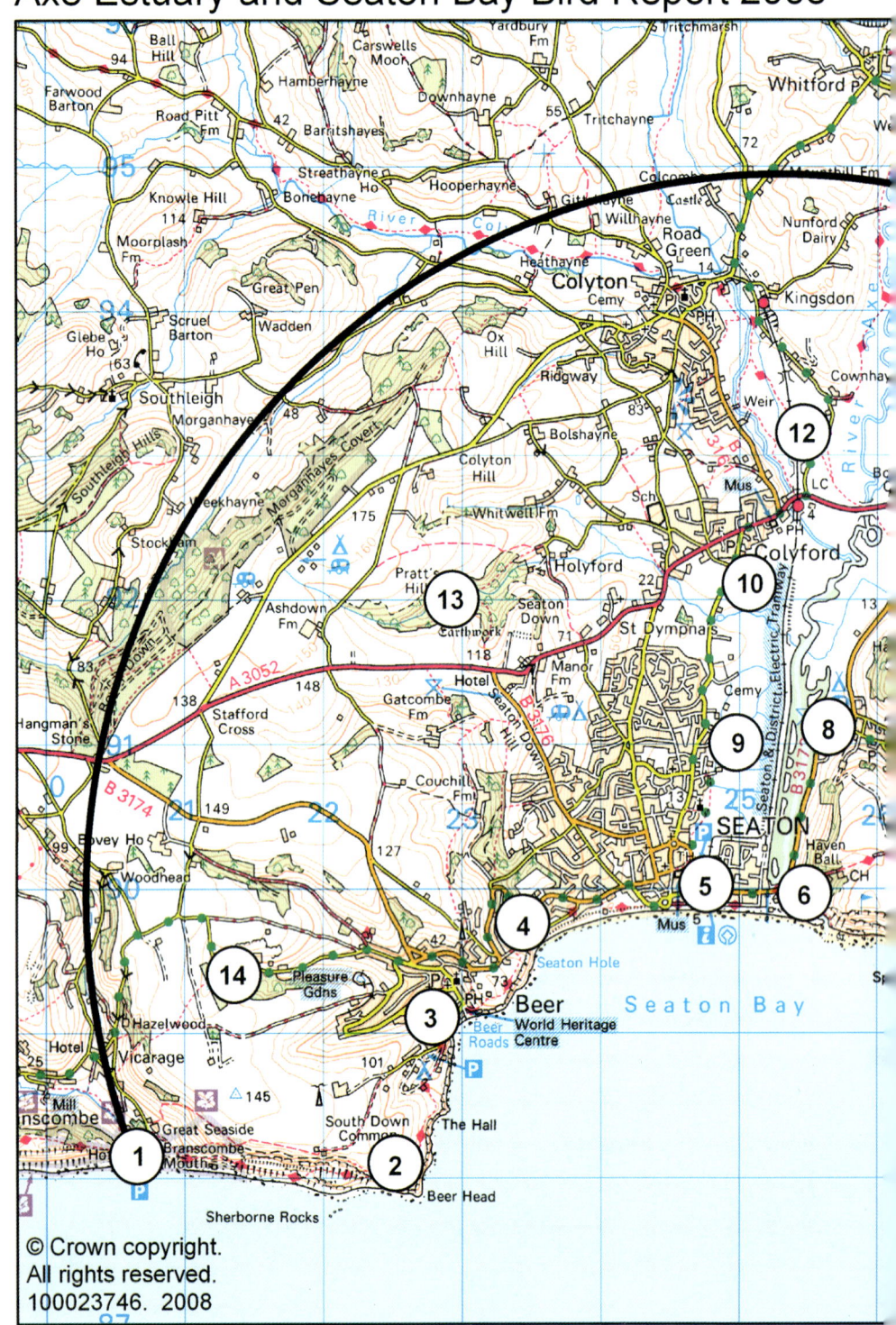

East Devon
District Council

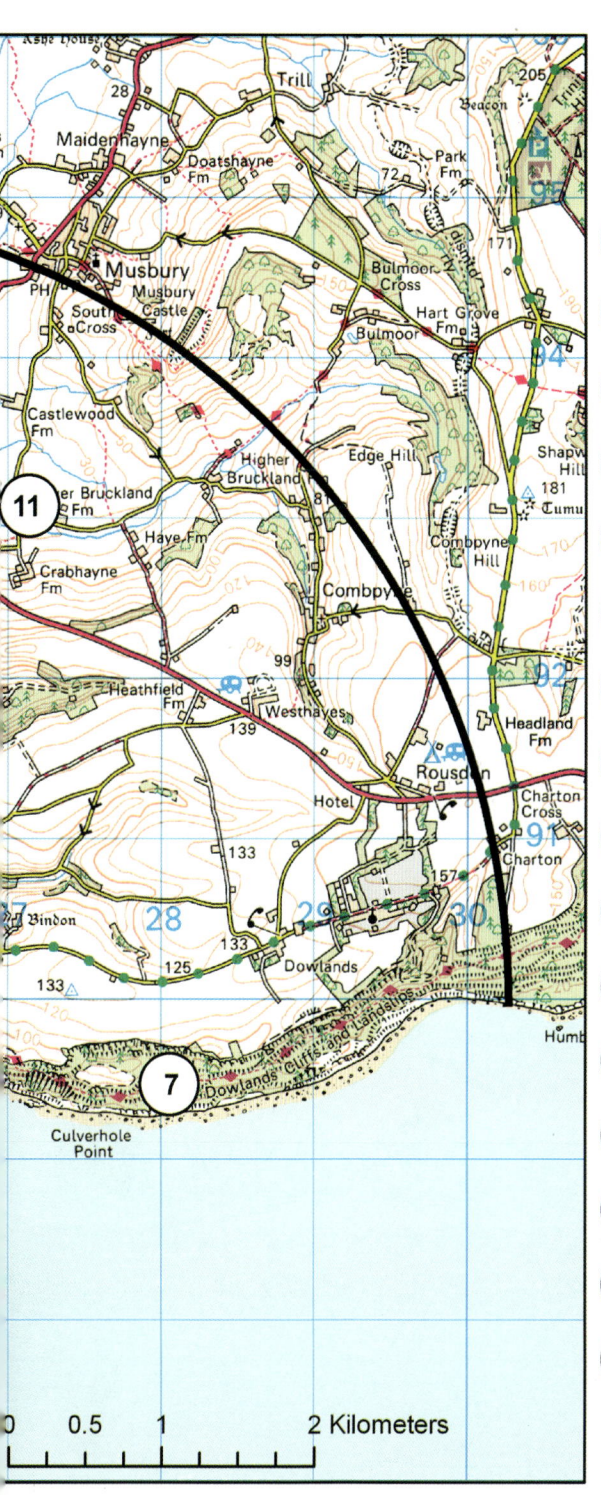

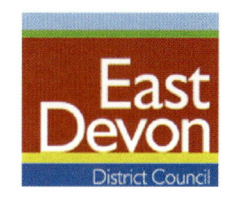

1. Branscombe Beach
2. Beer Head
3. Beer Beach
4. Seaton Hole
5. Seaton Beach
6. Axmouth Harbour
7. Axmouth to Lyme Regis Undercliffs National Nature Reserve
8. Coronation Corner
9. Seaton Marshes Local Nature Reserve
10. Colyford Common Local Nature Reserve
11. Lower Bruckland Ponds
12. Colyton Sewage Works
13. Holyford Woods Local Nature Reserve
14. Beer Quary
15. Boshill Cross
16. Haven Cliff

0   0.5   1   2 Kilometers

Axe Estuary 2008

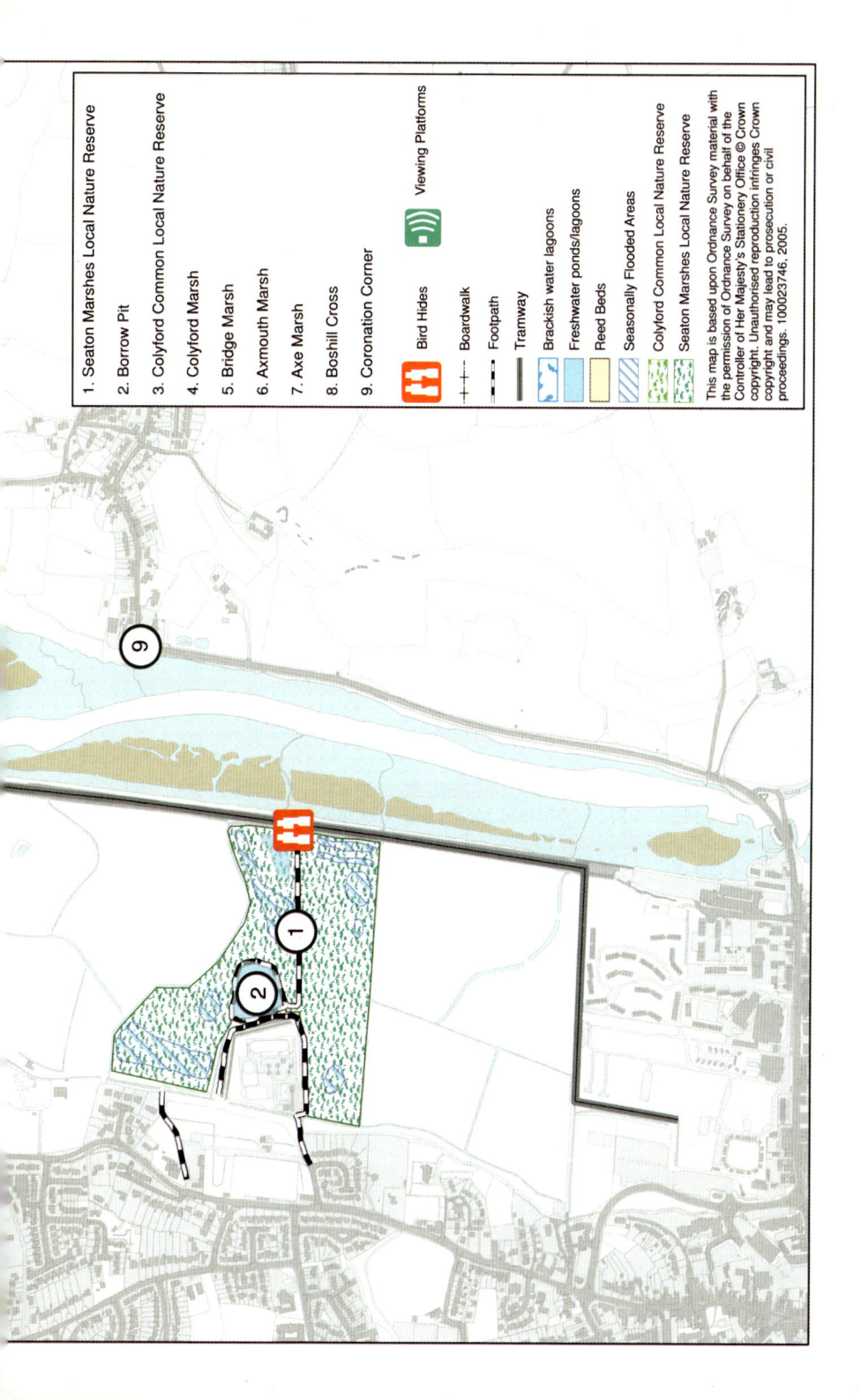

1. Seaton Marshes Local Nature Reserve
2. Borrow Pit
3. Colyford Common Local Nature Reserve
4. Colyford Marsh
5. Bridge Marsh
6. Axmouth Marsh
7. Axe Marsh
8. Boshill Cross
9. Coronation Corner

- Bird Hides
- Boardwalk
- Footpath
- Tramway
- Viewing Platforms
- Brackish water lagoons
- Freshwater ponds/lagoons
- Reed Beds
- Seasonally Flooded Areas
- Colyford Common Local Nature Reserve
- Seaton Marshes Local Nature Reserve

This map is based upon Ordnance Survey material with the permission of Ordnance Survey on behalf of the Controller of Her Majesty's Stationery Office © Crown copyright. Unauthorised reproduction infringes Crown copyright and may lead to prosecution or civil proceedings. 100023746, 2005.

The MSC Napoli Incident – Gavin Haig

On Thursday 18th January 2007 the 62,000-tonne container ship MSC Napoli sent out a distress signal after getting into difficulties in the English Channel. The vessel was taking in water through a hole in its side and the 26 crew were rescued by helicopter as the ship was battered in storm force winds some 50 miles off the Lizard Peninsula. The original plan had been to tow the ship to Portland Harbour, but salvage efforts were so seriously hampered by the weather and the Napoli so badly damaged that the Maritime and Coastguard Agency decided to run the ship aground on Sunday 21st January rather than risk it breaking up and sinking in Lyme Bay. Thus it came to be resting on a reef around a mile off the coast between Branscombe and Weston, on the edge of our recording area.

Around 200 of its 2,323 containers were washed into the sea, and the arrival of several of these on Branscombe Beach attracted massive media attention when they were found to contain BMW motorbikes and the like. This in turn prompted the arrival of scavenging 'gangs' from distant parts of the country, and a great deal of unsavoury behaviour. Though this was all meat and drink to the press there was a less spectacular, though arguably more sinister issue – 3,500 tonnes of fuel oil swilling around in the ship's undamaged tanks. Already, some 200 tonnes of oil had escaped from the engine room into the sea, creating an eight kilometre slick. The consequences were inevitable

Oil and birds do not mix
Oiled birds were noted from Sunday 21st January, and by 1st February the RSPB had received reports of 1,640 birds of 17 species from various locations around Lyme Bay, with particular concentrations at Portland and Prawle Point. Booms were floated on the lower Axe to prevent oil ingress into the river, but the noxious stench of it could be smelled wafting from the gulls gathered at Coronation Corner, where they endeavoured, in vain, to clean the muck from their plumage – quite pitiful. By around the end of February the RSPB had compiled the following table:

Species oiled	Number
auk sp.	62
Black-headed Gull	329
Black-throated Diver	1
Common Gull	31
Common Scoter	1
Cormorant	4
Great Black-backed Gull	16
Great Crested Grebe	1

The Axe Estuary and Seaton Bay Bird Report 2007

Species	Count
Guillemot	1105
gull sp.	102
Herring Gull	563
Iceland Gull	1
Kittiwake	1
Oystercatcher	1
Razorbill	30
Red-necked Grebe	1
Red-throated Diver	8
Ring-billed Gull	1
Shelduck	10
unknown	2
Total	2288

This total of oiled birds is the minimum count, and excludes both another 743 possible duplicates and data from the RSPCA. To give these figures a little perspective it is worth quoting some statistics from that organisation: during the week following the Napoli's grounding RSPCA animal collection officers and inspectors picked up 1,020 oiled birds between Torbay in Devon and Poole in Dorset. The majority came from the Portland end of the Chesil Beach area of Dorset. These comprised 967 Guillemots, 48 Razorbills, 3 Great Northern Divers, and singles of Red-throated Diver, Shag and Gannet. Of these, 467 Guillemots and 18 Razorbills were subsequently rehabilitated and released. The rest (more than half) died. The outlook for even mildly oiled birds that are NOT caught and cleaned is bleaker still, as the oil is ingested during preening and is extremely toxic. In addition, the oil removes the water-proofing from feathers, making it difficult, for auks especially, to dive, feed and keep warm.

Research has shown that the actual number of oiled birds following such an incident can be up to 10 times the recorded figures. With the wide distribution of casualties (indeed, many oiled auks were also found on the French coast in late January) it is possible that over 20,000 birds were affected.

In mid-July an attempt was made to refloat the Napoli. This was not successful, and was followed by its splitting in two using explosives. Despite the fact that the fuel oil tanks had been pumped out by early February some residual oil remained. During these operations it escaped, resulting in at least 140 more contaminated birds being noted, with a handful of mortalities. Once again, a boom had to be floated on the River Axe.

Ringing recoveries
Of 15 ringed birds recovered (mostly Guillemots) one was particularly unlucky – a Guillemot caught in the Erika oil spill off the coast of France in December 1999 had been successfully released in January 2000, but this time round it was not so fortunate and had to be put down. Even so, this bird does prove the immense value and efficacy of professionally cleaning oiled birds. Six of the ringed birds were from Great Saltee Island, off the coast of County Wexford in southeast Ireland, the oldest being a 21 year old Razorbill (now deceased). Others were from Sanda Island in west Scotland (including a 13 year old Guillemot), Skomer Island (Wales' largest breeding colony of Guillemots) and even as far afield as Fair Isle in Shetland.

The long term.
Post mortems carried out on 168 Guillemots and 104 Razorbills showed that the vast majority (80% or thereabouts) were adult birds (mostly males) and wing measurements suggested that they mainly originated from southerly breeding populations. Auks have low annual productivity, but can live for many years, so the loss of breeding-age adults is quite serious. Only time will indicate how serious.

Unexpected consequences?
It is said that every cloud has a silver lining, and this may have been true for the Napoli incident. There seemed to be a continuous presence of gulls downwind of the wreck, especially in rough weather. No doubt some of the containers below water level were leaching edible contents into the sea, creating an attractive 'chum' slick for the gulls. During the first winter/early spring period we enjoyed six Iceland Gulls (only two of which appeared prior to the Napoli wreck), 2 Glaucous, 2 Ring-billed and a Laughing Gull, as well as record day-counts of 749 Common and 12 Mediterranean Gulls on the estuary. It seems very possible that gulls following the coast would come across the convenient food source and stop awhile, then adjourn to the estuary for a wash and brush-up, perhaps lingering in the area much longer than might normally be the case. This is purely speculative but even if true, while it may appear to be some compensation for the ecological disaster of the wreck, what a horrendous price to pay for a few decent gulls.

Acknowledgements.
I am indebted to the following for various help that made this article possible: Kevin Rylands of the RSPB for oiled bird data, for pointing me to the post mortem information in BTO News (Sep/Oct 2007) and for invaluable comments on the draft of this article; also Rupert Griffiths, manager of the RSPCA West Hatch Wildlife Centre in Taunton, Somerset, for RSPCA figures.

At the time of writing (early 2008) the remains of the stern still lean precariously offshore. Salvage work is due to commence again in April to finally remove this blight from our coastline. It cannot be too soon.

Axe Estuary Ringing Group Report 2007
Mike Tyler (Group Leader)

Introduction
This is the first year since the establishment of the Group and has proved successful, with a wide variety of species caught and ringed. As usual mist and cannon-netting sessions were dependent on weather conditions, tides and bird movements. The objectives of the Group are to monitor and study the breeding, wintering and migrating patterns of birds along the Axe Estuary and, as a recognised group, to assist in the management of the estuary and act as a nucleus for cannon-netting sessions. This short version of the full report covers Group activities during 2007 at Colyford Common and Seaton Marshes Local Nature Reserves and on adjoining private land.

Ringing Activity
Species totals for the year are set out below showing a wide variety of species in comparison with previous years. Mist nets were set at Colyford in various habitats as in previous years; on land east of the tramline, including reed beds, opposite the hide, north and east of the ringing hut, and adjoining the Stafford Brook. A field close to the main road through Colyford was planted with various wild seeds, and was also included this year in the netting programme. Special effort was made to catch elusive pipits and this proved to be a partial success as can be seen from the list below. One cannon-netting session and two evening wader catches took place. A total of 491 birds was caught at Colyford, of which 48 were retraps/controls.

At Seaton Marshes three cannon-netting sessions took place to trap and ring wildfowl, and during the spring morning sessions for migrants took place at the Borrow Pit. Early winter saw the arrival of two Abberton Traps used for the trapping of wildfowl. A total of 311 birds was caught, of which 27 were retraps/controls.

These activities resulting in a total of 802 birds caught and processed of which 73 were retraps and 2 controls (birds ringed elsewhere).

Future Activities
A programme for trapping and ringing birds during 2008 within the area of the Axe Estuary has been prepared and covers all months of the year and a wide selection of habitats. New methods of trapping are envisaged, including the use of various traps and whoosh nets that have been tested and hopefully will prove useful during the year. The use of tape lures outside the breeding season will again be used as this method proved helpful in previous years. The possibility of catching gulls is being considered on sites close to the water treatment works, as well as waders and wildfowl on the new islands and lagoons created by East Devon District Council Countryside Service, both at Seaton and Colyford. Cannon-netting sessions will also continue in order to catch large numbers of wildfowl.

The Axe Estuary and Seaton Bay Bird Report 2007

Acknowledgements

The Group would like to thank Peter Dare, Seaton and District Electric Tramway Company and East Devon District Council for allowing access to their land. Grateful thanks are given to East Devon District Council, Area of Outstanding Natural Beauty (Community Fund), Devon Birdwatching and Preservation Society, Axe Vale and District Conservation Society and Seaton Lions Club for their generous funding. Thanks also to Robin Ward and Richard Hearn of the Wildfowl and Wetlands Trust, members of the Group, several visiting licensed ringers, group leaders and helpers who joined us during the year and made the sessions a success.

Bird Ringing at Colyford Marsh *(Photo by Roger Boswell)*

List of species trapped and ringed (numbers in brackets refer to retraps/controls) :

Colyford

Shelduck 5, Kestrel 1, Oystercatcher 4(1), Little Stint 1, Dunlin 3, Green Sandpiper 1, Woodpigeon 1, Kingfisher 8(5), Great Spotted Woodpecker 2, Sand Martin 1, Swallow 10, House Martin 4, Meadow Pipit 15, Rock Pipit 1, Water Pipit 3, Wren 9(5), Dunnock 50(11), Robin 23(7), Stonechat 6(2), Blackbird 25(2), Song Thrush 19, Redwing 3, Cetti's Warbler 5(3), Sedge Warbler 9, Reed Warbler 17(5), Blackcap 2, Chiffchaff 24, Willow Warbler 2, Goldcrest 4, Long-tailed Tit 18 (4), Blue Tit 27(2), Great Tit 5, House Sparrow 11, Chaffinch 79, Greenfinch 2, Goldfinch 36, Linnet 2, Reed Bunting 4(1).

Total 443(48)

Seaton

Shelduck 54(16), Wigeon 59(1), Teal 3, Mallard 79(1), Moorhen 11, Wren 4(1), Dunnock 4, Blackbird 4(2), Song Thrush1(1), Cetti's Warbler 1(2), Blackcap 1, Chiffchaff 3, Willow Warbler 2, Blue Tit 2(1), Great Tit 7, Carrion Crow 31(1), House Sparrow 5, Chaffinch 3(1), Greenfinch 8, Goldfinch 1, Reed Bunting 1.

Total 284(27)

Overall Total 727(75)

Dragonflies and Damselflies (*Odonata*) – Steve Waite

An increase in man-hours in the field has not only seen an increase in bird sightings/records, but other fauna records have benefited from this also, with the status of several dragonfly species changing, both on a local and sometimes county scale. Since **Red-veined Darters** *(Sympetrum fonscolombii)* and a single male **Lesser Emperor** *(Anax parthenope)* were found at Lower Bruckland Ponds (LBP) in 2006, this site has proved to be one of Devon's most impressive dragonfly hotspots, with the county's first **Small Red-eyed Damselfly** *(Erythromma viridulum)* found here also in 2006 – with 54 the peak count by the end of the season! 2007 did not disappoint, highlights included both **Small Red-eyed Damselflies** and **Red-veined Darters** still at LBP, and the first **Scarce Chasers** *(Libellula fulva)* for the area. Here is a brief account of Odonata sightings on patch in 2007, with some comments on local status and distribution:

Beautiful Demoiselle
Calopteryx virgo
Prefers smaller and quicker moving areas of water than *C.splendens*, with the two most reliable sites being below the small foot bridge at LBP and along the Stafford Brook at Colyford Common. Otherwise common in suitable habitat.

Beautiful Demoiselle *by Mike Hughes*

Banded Demoiselle *Calopteryx splendens*
Abundant, largest count of the year being 400+ beside the Axe by the A3052 road bridge on June 8th.

White-legged Damselfly *Platycnemis pennipes*
Abundant along the Axe, especially near the confluence of the Coly and at Whitford. Mass emergence took place in early June, with 450+ along the river banks near the A3052 road bridge on one morning.

Large Red Damselfly *Pyrrhosoma nymphula*
Common in places, as most often is the case the first Odonata of the year on the wing with one at Seaton Marshes on Apr 5th.

The Axe Estuary and Seaton Bay Bird Report 2007

Red-eyed Damselfly *Erythromma najas*
For the second successive year small numbers present at LBP early-mid Summer, highest count made on June 5th with four, no mating observed.

Small Red-eyed Damselfly *Erythromma viridulum*
For the second successive year present in good numbers at LBP, although the peak count was only half of last year's best, with 30 on July 24th – this may have been due to a lack of surface weed on several of the ponds or because of poor weather during their flight period, or both! The first sighting of the year was of three on July 8th. As with last year, mating and ovipositing observed. The species was first found here in 2006, when it was the first record for Devon and this remains the most westerly site in the UK for this species.

Azure Damselfly *Coenagrion puella*
Abundant during early and mid summer preferring slower moving water, particularly common in the ditches at Seaton Marshes.

Common Blue Damselfly *Enallagma cyathigerum*
Same as *C.puella*, though flight period tends to last longer - this year large numbers were still on the wing in mid-September. Particularly common at the Seaton Marshes Borrow Pit

Blue-tailed Damselfly *Ischnura elegans*
Abundant throughout mid summer around most types of water (rivers, marshes, large ponds, garden ponds). Seaton Marshes and LBP support large numbers in particular.

Migrant Hawker *Aeshna mixta*
Not an outstanding year for this species, though still seen in relatively good numbers from early August, often some distance from water.

Southern Hawker *Aeshna cyanaea*
Regularly encountered, though never numerously, from August, with specimens still on the wing in October. Like *A.mixta* often seen patrolling in gardens and along pavements!

Emperor Dragonfly *Anax imperator*
Common, present at any suitable habitat, biggest counts coming from LBP. First sighting of year May 24th (six).

Golden-ringed Dragonfly *Cordulegaster boltonii*
Sparsely distributed throughout the patch, often away from water (which may be why relatively few are recorded?).

Four-spotted Chaser
Libellula quadrimaculata
Only recorded at LBP, with a maximum of three present in mid July.

Four-spotted Chaser

Scarce Chaser *Libellula fulva*
The first record of this species for the patch was a teneral at LBP on June 6th, but on 8th at least six were seen between the confluence of the Coly and Whitford along the Axe; they were seen here regularly for the following couple of weeks. Occasional individuals were seen at LBP throughout June, though at least three were present here on July 8th, a female and males with mating scars.

Broad-bodied Chaser *Libellula depressa*
Seen throughout the patch during mid-summer, though never numerous, only found in ones and twos, with both the marshes and LBP being favoured spots.

Black-tailed Skimmer *Orthetrum cancellatum*
Most abundant at LBP, where very obvious especially during June and July. Lots of mating and ovipositing observed. Also common around the deep lagoon at Seaton Marshes

Keeled Skimmer *Orthetrum coerulescens*
One at LBP in early Aug was the only record of the year on patch, although data from previous years suggest this must only be due to under-recording, and not going to the right sites at the right time!

Common Darter *Sympetrum striolatum*
Probably the most abundant Odonata of all, with the longest flight period. Present throughout the patch in gardens, as well as being present in large numbers at the usual Dragonfly hotspots. First recorded on June 4th, with individuals still being seen in mid-November.

Red-veined Darter *Sympetrum fonscolombii*
Present at LBP for the second successive year, though they preferred a different pond this year. First sighting was of a teneral/female on June 4th, then at least seven (including a pair watched egg laying) were seen during the next few days. Then followed a gap in sightings till late July, when a couple of mature males were expected to be the last sightings of the year. But on Sept 16th a freshly emerged teneral (and its exuvia – the first to be found in Devon for this species) were found by the pond favoured by this species earlier in the year. The final record was of a female-type on Oct 6th.

Ruddy Darter *Sympetrum sanguineum*
Recorded from four sites within the patch boundary from mid-summer onwards, with Seaton Marshes remaining the prime site. Also seen occasionally at LBP and Colyford Common, plus a single record of an immature male in Beer Quarry on July 9th.

Ruddy Darter

I would like to thank everyone whose records have helped to compile this account, in particular Dave Smallshire who spent a lot of time around the Axe this year and is responsible for many of these sightings.

Butterflies – Steve Waite

Unlike *Odonata* only the rare/scarce/unusual species have been listed here with details of their sightings. Small Skipper, Large Skipper, Brimstone, Large White, Small White, Green-veined White, Orange Tip, Small Copper, Common Blue, Holly Blue, Red Admiral, Painted Lady, Small Tortoiseshell, Peacock, Comma, Speckled Wood, Marbled White, Gatekeeper, Meadow Brown and Ringlet can all be found with relative ease in the right habitat during their flight periods. Here are accounts of the scarcer species that were seen on patch in 2007, but please note that a few sites are sensitive as they may be private or liable to disturbance, so some locations are vague.

Dingy Skipper *Erynnis tages*
Still present in two areas on patch, on the western edge of the area no more than three recorded in late August, but near Axmouth there was an impressive series of records. Early records came on Apr 17th with 6-10 on the wing. In the late summer period up to twenty (probably well over) were recorded at the same site.

Dingy Skipper *by Steve Waite*

Grizzled Skipper *Pyrgus malvae*
A true rarity on patch, but one was observed on Beer Head on Apr 18th.

Wood White *Leptidea sinapis*
Two sites on patch support this species. Records from the Branscombe area show two flight periods, with sightings of up to three between Apr 29th and May 4th, then again between Jul 14th and Aug 24th with a peak of three. Only other records on patch were from Undercliff NNR with four the peak count on Aug 4th.

Clouded Yellow *Colias croceus*
Some interesting records this year for this species, with some very early records from the Undercliff NNR possibly suggesting overwintering. Sparse throughout the summer with little sign of migration, though a small arrival was evident in September. Other unusual records from the Undercliff NNR include two fresh individuals on Nov 1st with one on Nov 9th.

Green Hairstreak *Callophrys rubi*
Two sites on patch support this species (as well as the as yet unfound sites!). Beer Head appears to support the greatest numbers with up to six here in early summer; also two records of singles in Undercliff NNR on May 3rd and Jun 6th.

Brown Argus Piebeius agestis
From first brood the only record was three on Undercliff NNR on May 5th. Then recorded between Jul 19th and Aug 24th from Undercliff NNR or near Branscombe, with the largest count of five at the latter on Aug 5th.

Chalkhill Blue Polyommatus coridon
Seen between Aug 3rd and Aug 28th on the western edge of the patch with the highest recorded count being four on 13th though 'several' were noted here on Aug 4th.

Chalkhill Blue by Steve Waite

Large Tortoiseshell
Nymphalis polychloros
The rarity highlight of the year with a fresh individual observed on Musbury Castle on Jun 20th. This sighting was one of many along the south coast of Britain for this species this year (Portland clocked over a dozen records!). This was perhaps one of the most unusual occurrences in the natural world in the UK in 2007!

Large Tortoiseshell by Mike Lock

Dark Green Fritillary Argynnis aglaja
One was seen in flight near Branscombe on Aug 5th.

Silver-washed Fritillary Argynnis paphia
Sporadic sightings throughout the patch in July and August with a peak of three at Undercliff NNR on Aug 4th.

Wall Lasiommata megera
Peak count eight near Branscombe on Aug 3rd; several present on Beer Head and along Undercliff NNR also in mid to late summer.

Thanks must go to Phil Parr of Butterfly Conservation - Devon Branch, whose records were vital in the production of this article, also thanks to Mike Lock and Donald Campbell for their records.

The Axe Estuary and Seaton Bay Bird Report 2007

Audouin's Gull at Seaton Marshes – Gavin Haig

Around the turn of the year a few of us had been speculating about potential 'firsts' for the patch. Our guesses mostly involved fairly predictable species. However, reality can be far more adventurous, and so it proved when my phone rang just before 11:00 on Tuesday 14th August. Without preamble I heard Steve Waite's slightly tense voice announce "there's an Audouin's Gull at Seaton Marshes!" "Really?" was my pathetic response. Extremely thankful that I was nearby, I joined Steve in the hide a few minutes later. A flock of mostly big gulls was visible NW of the hide, in a field at least 500m distant. They were evidently taking shelter from the strong and drizzly southerly wind. Between the gulls and us was a field of rank weedy growth, plus a barbed wire fence festooned with bramble. Consequently only a fraction of the gull flock was fully visible, and even those few birds on view managed to obscure one another to a great extent. Remarkably, when Steve had first scanned the flock the Audouin's Gull had been in full view! But now it was not. I could see the back of a head, a bit of mantle, and some wing tips, but almost immediately the bird became totally obscured. Great. Reluctantly, Steve needed to leave on an errand and would have to get back later. I agreed to keep my eye glued to the spot where the bird had just disappeared until other birders arrived. After several minutes alone I suddenly realised that I could see it – a pure white, beady-eyed head was visible, sporting a stonking red bill! I was absolutely stunned! Gradually other birders began to arrive, but the bird was extremely difficult to see, more often than not partially or completely obscured by other gulls, fence posts or bits of vegetation! After more than an hour of brief and tantalising glimpses the bird finally stood in the open for about ten minutes, preening, allowing me to take several digiscoped photographs. When Steve did eventually return he was able to view the flock from a house overlooking the marshes from the west, and captured a blurry but clinching shot himself. I had to leave the hide around 13:30 and a little later the flock dispersed, the Audouin's disappearing. We guessed it had headed out to sea, but there was to be an unexpected twist in the tale…

Description
Size – an oddly sized gull, smaller than Herring, but not a great deal smaller than Lesser Black-backed.
Structure – mostly hunched in the poor weather, but occasionally stretched its neck, when it looked somewhat sleeker. Fairly hefty bill, and the bill feathering (particularly on the upper mandible) was quite extensive, giving the bird a long sloping forehead and 'snouty' appearance when seen side-on.
Plumage – uniform pale grey upperparts (mantle, scapulars and coverts), fairly similar to, or paler than nearby Herring Gulls. There were no obvious white tips to the tertials, ie. a 'tertial crescent' – if it had one it was not striking. A little unevenness along the edges of the coverts suggested some wear. The underparts had a silky smooth appearance – rather different to the other gulls, almost as if it had finer, less coarse feathering – and were washed in a very

pale grey, to upper breast level. This grey wash was subtle, but quite striking and unusual looking. It looked darker on the belly, but this may have been an effect of shadow. In contrast, the head was a stark, bright white, extremely clean. What was visible on the closed wing indicated that the primaries were black. No white at all was visible on the closed wing, the primary tips appearing entirely black.

Flight views – the bird was seen at least three times in flight, as it changed position in the flock. It appeared that the only black in the wing was on the primaries. It was quite extensive, at least as much or perhaps more than on an adult Herring or Common Gull, and similarly shaped, but lacking any obvious white – either tips or mirrors. Otherwise the wing looked entirely grey. The tail appeared to be unmarked white.

Bare parts – the eye was small and looked dark. Combined with the clean, white head this gave the bird its 'beady-eyed' appearance. The bill was strikingly red, with a darker, possibly blackish distal third. It was not a bright red, more a blood red, but extremely noticeable, even at the long range and in the dull light. Bill structure was quite long and heavy-ish, certainly would not be described as dainty! The legs were darkish, probably grey.

Ageing – everything fitted adult, except:

1. It is just possible that there may have been a very small amount of black elsewhere on the wing (primary coverts, for example) that would age the bird as a third summer. However, no such markings were discernible at the range involved.
2. Lack of white primary tips, suggesting immaturity. However, there are photos of adults on the internet, taken in late July and August, where all the white tips are missing – worn off. In the photograph on page 45 the longest primary is clearly blunt and worn. Quite likely all the other primaries were in a similar state.

In conclusion then, it was probably an adult, but may possibly have been a sub-adult, ie. third-summer.

The Twist

No further confirmed sightings of the bird were made that afternoon. However, shortly before 21:00 Dan Pointon and Stuart Piner, who had endeavoured to twitch the gull from the Midlands, had given up the search and were seeking a local supermarket. Spying the rather utilitarian roof of the Lyme Bay Holiday Village they surmised that they had chanced upon a Tesco store, or similar. They stopped and got out of the car, Dan casually passing his binoculars over the few gulls perched on the roof. Unbelievably the Audouin's Gull was amongst them! They were able to take the best photographs of the day! Unfortunately, at 21:08 the bird headed off over rooftops towards the sea, just before other observers arrived on the scene.

Discussion

As recently as the early 1970s Audouin's Gull was a globally threatened species, with fewer than 1,000 breeding pairs. A dramatic improvement in fortunes has led to a current population of around 19,000 pairs, with 17,000 of these in Spain (ref. 'British Birds' vol.100 p.49). Even so, it is still a very rare bird north of its Mediterranean basin breeding range. This record constitutes the fourth for the UK (the first was in May 2003), and the first for Devon. The previous three were all second-summer birds, so this one breaks the mould.

Bonaparte's Gull on the River Axe – Steve Waite

The last day of April was a lovely sunny day, and where better to be than bird ringing in a large garden in Combpyne! I left here just prior to 11am, and on the way home, decided a scan through the gulls on the estuary would be worthwhile...

I pulled in at Coronation Corner and could see a large flock of gulls upriver. Nearly all were large gulls – and I felt a scan through with my bins was sufficient, but when I noticed a huddle of about thirty Black-headed Gulls further upriver the telescope just had to come out! A strong heat haze made viewing difficult, but I could see enough on one particular gull to set alarm bells ringing! In with these Black-headed Gulls was a first-summer gull which gave the appearance of looking somewhat duskier around its head – reminding me a bit of a Franklin's Gull – and despite looking hard I just couldn't make out anything other then black on its bill. These features, along with its apparent slightly smaller size were enough to indicate to me that this may be a Bonaparte's Gull! I rang Gavin Haig who immediately set off, but before he arrived the bird walked out of the water revealing a pair of pink legs, this really got my heart pounding! About six minutes after I phoned him, GH set his scope up next to mine and immediately agreed that it did indeed look good, but he wanted to get a tad closer. As he wandered on to Coronation Corner picnic site a low flying helicopter spooked all the gulls. Luckily, I was watching the gull when it flew, and as it took to the air it revealed lovely pale under wings – and it really did look small and compact, similar to a Little Gull! These views prompted me to run over to GH exclaiming "it is one, it is one!!!". It soon landed again, a lot closer in the middle of the river just opposite the picnic site, there it was, a wonderful first summer Bonaparte's Gull! I returned to the car to get my camera, but before I had time to attach the camera to the telescope it took flight and started heading off high northwest with one of its commoner cousins. Much to our frustration it kept flying, and we eventually lost it over a distant hill. We didn't know how to feel, jubilant in the fact we had just scored with a highly desired patch first, yet dismayed that it had already gone!

I rushed home to start writing my description, but ventured out again mid afternoon for another look along the estuary. At about half three I had reached Coronation Corner and one could only imagine my utter thrill at relocating the bird sat beside the waters edge alongside forty or so Black-headed Gulls . My mobile phone soon kicked into action and it wasn't long till the local birders started arriving. It showed well for the next hour or so allowing the cameras to click away, but I had to leave it to return home at 5pm. A look again later in the evening after the tide has risen failed to locate it, although an Iceland Gull opposite Coronation Corner made the trip worthwhile!

This small North American gull has never occurred on patch before, despite records from the Otter, Exe and several other south coast localities. For me it was one of my highlights of the year, to find a subtly different gull is always a great challenge, and it is why I spend several hours a week scanning through large flocks of gulls!

Iberian Chiffchaff at Beer Head – Steve Waite

By late morning on Saturday, April 28th I was kitted up and ready to go birding. It was quite a clear day, so I thought best to go on high ground if I wanted any chance of bumping in to some migrants. So, I opted to give Beer Head a visit – our local migration hot spot! I took the usual route and headed towards the viewpoint that looks down into the undercliff. So far, so poor – no migrants in sight! I got to the viewpoint at about 11am and immediately heard a striking burst of song from a bird nearby in cliff-side bushes, it sang again shortly after and it struck me it sounded like an Iberian Chiffchaff, a species I have seen once before in Yorkshire. It sang another couple of times, but then that was it, not a squeak - I looked and looked, but no sign. Half an hour had passed and there was no further sound – let alone sight! I really thought that was it, the one that got away! I hadn't even seen it, but still I thought it necessary to send a text out to fellow local patch birders which read "Chiffchaff singing with several Iberian qualities at Beer Head". I then left Beer Head to see a cracking drake Garganey which had been found down on Seaton Marshes. Here I met fellow local birder Phil Abbott – we spoke a bit about the warbler and agreed to go back up there after lunch.

At about 14:00 I took PA to the spot and was amazed to hear it sing virtually as soon as we got there. I spent the next three and half hours with the bird, Gavin Haig, Ian McLean and Karen Woolley all joining me through the afternoon. It sang virtually the whole time, although it would remain quiet for the odd 5-10 minute period. The more I heard it the better it sounded, we even heard it call a couple of times. These are the notes I sent to BBRC about its song/call:

Song
A very distinctive loud and powerful song, really ear-catching and reminded me of the Yorkshire bird – hence why I realised from the start that this was probably an Iberian! There were three parts, but on some occasions the bird would sing just the first two parts, and a couple of times it would miss the middle part out. The final section had a very Cetti's Warbler-like feel to it, both in the actual sound, and also in the rather explosive way it was delivered, this section was delivered at a quicker pace than the first two. At no time did it go Chiff Chaff, not even Chiff or Chaff – despite

the presence of 2-3 singing male Common Chiffchaffs further down in the undercliff. I would transcribe its full three-phrase song like this... 'chu, chu, chu, weep, weep, che-che che che'

Call
Unlike its song, it only called on three occasions, all in quick succession. This was heard by myself and Gavin Haig. We were both surprised at how unlike a Chiffchaff or Willow Warbler it sounded, we both reacted similarly by saying "wow – that's different!". It was a down-slurred rather plaintive 'teu', quite a soft call too – not piercing like the song.

Whilst listening to the bird, despite hearing it sing for such a long period we knew we needed some sound recordings of it. GH and myself both tried recording with our mobile phone, but we both failed miserably! We needed help - which is when Karen Woolley became the hero of the day, armed with two camera/camcorders with sound recording capacity - great! She spent the next hour trying her all to get some good recordings, but it wasn't till the following day that we found out just how good a job she did with some excellent sound clips.

So we heard it enough, but what about actually seeing it?! The occasional views it did offer were always brief, and mostly in the shade of bushes and shrubs, but after several hours of these odd glimpses the notes on plumage and behavior built up and enabled me to write the following in my description:

Plumage and Behavior
The bird was very elusive – not giving itself up for prolonged clear views at any point during the observation, this is probably partly due to the thick cover it was in, but I felt it was behaving slightly different too. I'm sure a Common Chiffchaff or Willow Warbler would have spent at least some time perched up fly-catching from the bush tops – this bird behaved more like a skulky Sylvia or Hippolais warbler! I remember noting this feature with the Yorkshire bird.
Despite its elusiveness, as I spent the most amount of time with the bird I had the best chances to study it. These are the points I noticed with the relative poor views I had. Note that all views of it were in the shade of the vegetation.
It basically looked like a Common Chiffchaff apart from a few odd features, the most striking being its bright olive green rump which contrasted strongly with its browner mantle. Other features include:
Neat yellow supercillium extending behind eye with dark eye-stripe, also yellow-tinged ear coverts.
Pink feet and lower legs (didn't see above the joint well enough to note its colour).
The bill looked mostly pale.
On a brief glimpse from the rear the primary projection appeared to be on the long side for a Chiffchaff, approaching more Willow Warbler-like.
The bird also looked unusually white-bellied for a Chiffchaff – giving the appearance more of a Garden Warbler!

And that was it! I left Beer Head late afternoon, and it was never seen/heard again. So, we knew what had to be done. Fingers went to keypads (gone are the days of pen to paper!), sound clips were listened too and GH had the initiative to ask for the opinions of the experts. The experts duly replied and the feedback was all positive, with sonograms confirming the pitch of the song notes were spot on for Iberian Chiffchaff, apparently ruling out 'dodgy' Common Chiffchaff or hybrid. I would like to thank to Martin Collinson and Magnus Robb for their responses.

This is the first Iberian Chiffchaff for the patch, the fourth for Devon and only about the fifteenth for Britain.

Postscript: This individual featured in an article entitled 'Identification of Vagrant Iberian Chiffchaffs' in 'British Birds' vol.101 pp.186 and 186

Stone-curlew at Seaton Marshes – Steve Waite

A text message from Ian McLean woke me up on 13th April to say he had seen singles of Pied Flycatcher and Ring Ouzel in the Underhooken below Beer Head; two quality spring birds for the patch – it was looking promising for the day. A glance out the window showed it was a foggy and very damp morning, which for me means go to low ground. I find Beer Head is best with clear skies as birds are passing higher in the sky, but when the weather is lousy and damp, migrating birds seem to keep low in the river valley, hence the reason the quality birds at Beer Head were in the Underhooken below. So Seaton Marshes would be my port of call, though I couldn't get there till 10 am due to other commitments (being a taxi!).

I rolled into the car park just before ten, and it looked immediately promising, with several Swallows flying low over the scrapes, but a walk around the Borrow Pit revealed next to nothing in the bushes. I headed up the path towards the hide, pausing by the first gap in the screen to peer through towards the dried out scrape. My naked eye picked out a brown blob on the grass just in front of the (water-less!) scrape, which could easily have been a heap of dried earth, or a stone! But for some reason a message was going to my brain to 'put my bins on to it!', so I did, and it nearly sent me into a state of shock! It was the front half of a Stone-curlew!! Probably one of the biggest birding shocks my body has ever had to deal with, not so much for the rarity value – just the sheer shock of a Stone-curlew being sat there in front of me on Seaton Marshes as though it was 'the norm'!

I fumbled about with my mobile and phoned all the local birders and a few county birders, but before I finished my last call the worst thing imaginable happened (and seems to happen far too often with scarce birds!) – it got up and flew! Part of me was almost mesmerized with this Stone-curlew flying around low in front of me, but most of me was thinking "Oh no! Where's it going?!" Thankfully it banked around and landed in the field to the south of the marshes in full view, phew! It wasn't long before a phone call from Gavin came through to say he was on site, my reply was something like "run run run!!" which I believe he did, but it turned out there was no necessity as it stayed on view in this field for the next fifteen minutes or so. Subsequently it walked into a ditch, which is when I had to leave, but the bird remained in the area for the rest of the day. At first it stayed on Seaton Marshes but when the tide dropped it flew onto the estuary where it fed beside the river bank north of Coronation Corner till dusk. I saw it here later in the day.

It turned out to be a very good day for birds in the valley. After I found the Stone-curlew I saw singles of Little Ringed Plover and Goosander fly over, and birders watching the Stone-curlew mid afternoon along the river were treated to an adult Little Gull and a couple more Little Ringed Plovers. It just shows the right conditions really can do the trick!

This is the second Stone-curlew on the patch following one at Colyford Marsh as long ago as April 1963! Despite recent records for Devon, this was the first twitchable Stone-curlew in the county for at least twenty years, hence it was a very popular bird, with at least 60 birders making the trip to see it. Most interesting though was the fact this bird was colour ringed, with the colour combination being read in the field. A quick response confirmed that this bird (ET44595) was ringed as a chick in Elveden near Thetford, in Norfolk's Breckland, on May 27th 2005. It and its sibling were seen to fledge, but unlike its sibling it hadn't been reported since. Of even more interest is that this is the furthest west that a Norfolk-ringed Stone-curlew has ever been reported.

The Axe Estuary and Seaton Bay Bird Report 2007

East Devon District Council Local Nature Reserves – Fraser Rush

Within the recording area there are three Local Nature Reserves (LNRs) owned by East Devon District Council. The title 'Local Nature Reserve' comes from an act of parliament dating back to 1949 which provides a legal framework for declaring sites which are good for both wildlife and the visiting public. Typically (though not always) managed by local authorities, there are over a thousand LNRs in England which range in size from entire estuaries to individual trees.

There are currently eight LNRs in East Devon, more than any other district in Devon, and covering habitats as diverse as heathlands, coastal marshes, woodlands, sand dunes, inter-tidal mud and flower-rich meadows.

There follows a description of the three LNRs within the recording area together with brief details of the facilities available at each one. Full access and location details for each site can be found on the East Devon District Council website:
www.eastdevon.gov.uk/local_nature_reserves

Seaton Marshes LNR

Having been in Council ownership for many years, this 10.5 ha (26 acre) grazing marsh was declared a Local Nature Reserve in 1999. Seaton Marshes lies within the area that was reclaimed from tidal saltings in the seventeenth century. The LNR comprises two fields, separated by a modern flood bank and known originally as Little Ragged Jack and Great Ragged Jack.

Seaton Marsh Bird Hide (Photo by Fraser Rush)

Habitat and Management

Since 1999 East Devon District Council has implemented a series of hydrological works, aimed at creating better conditions for wildlife, particularly wintering birds. A number of new earth bunds, pipes and sluices allow large parts of the marsh to flood during the winter months, creating shallow water favoured by a number of wildfowl and waders. Various scrapes provide slightly deeper winter flooding and one deep lagoon holds water throughout the year. Most of the flooded areas are allowed to dry out gradually in the Spring, leaving them completely dry by mid-summer.

The timing of the flooding regime, combined with summer grazing by sheep and cattle, creates perfect conditions for many birds in the winter. At peak times the LNR can hold more than 250 Wigeon, 100 Teal and 100 Shelduck, together with smaller numbers of Shoveler, Black-tailed Godwit, Snipe, Curlew and a wide range of other winter visitors.

The boundary ditches around the fields are managed, not only for their hydrological efficiency, but also to maximise their wildlife value. They hold good populations of dragonflies, including the Ruddy Darter which has its local stronghold here. The ditches also support populations of fish, notably roach and eels, the later providing good food for otters which are frequently present (though seldom seen).

The Borrow Pit

Another interesting area is the Borrow Pit nature reserve. Surrounded by Seaton Marshes on three sides, the Borrow Pit was dug in the early 1980s to create the large flood bank across the marshes. It has been managed as a nature reserve for many years by the Axe Vale and District Conservation Society and is now part of the LNR.

As the only large area of deep water, the Borrow Pit supports a number of species not found elsewhere in the area. Little Grebe usually breed here and Mute Swan have done in the past. Little Egrets frequently roost in the trees on the island and a number of warblers can be found in the surrounding scrub and emergent vegetation, particularly during the spring passage.

Access and Facilities

Visitor facilities at Seaton Marshes have improved considerably since the LNR was declared. A screened, wheelchair friendly path leads to a large hexagonal bird hide which overlooks both the tidal estuary and the eastern half of the LNR. Many birds can be seen from this hide, often at relatively close range. A further path leads around the Borrow Pit, a particularly pleasant spot during the summer, and another goes around the adjacent sewage treatment works, giving views over the northwest corner of the LNR.

Colyford Common LNR

A diverse area of marshland at Colyford Common was notified as a Local Nature Reserve in 2002. Comprising land acquired by the District Council as well as land belonging to the Mayor and Burgesses of Colyford, the LNR totals 13.5 ha (33 acres).

Colyford Marsh viewed from Colyford Common bird hide (Photo by Fraser Rush)

Habitat and Management

In contrast to Seaton Marshes, most of Colyford Common is subject to tidal inundation, although it lies at the very top of the tidal gradient, the amount of land covered by the incoming tide is highly variable. On a few occasions every year, usually in late autumn or winter, three quarters of the LNR is flooded, whereas on most tides the only noticeable effect is deeper water in the ditches. What makes the area very special is that the area of land flooded by the tide is not restricted by man-made sea defences. Instead each tide finds its own height relative to the natural topography.

The tidal areas of the LNR are grazed by cattle during the summer. The herd which presently graze the site are Devon Red Rubies, specifically managed for conservation grazing. The farmer involves local volunteers in managing the herd and markets the beef locally – for details, see the notice boards at the nature reserve.

A few areas of the nature reserve remain above the height of even the highest tides and these areas are managed in a variety of ways to encourage a greater diversity of wildlife. To the north of the tidal marsh there is a wet field which has been made even wetter by alterations

to the way it drains into the boundary ditches. The field has been planted with clumps of reed which will gradually spread to cover the entire area, favouring the wetter conditions which have been created and ultimately forming a dense reed bed. At the time of writing (Winter 07/08) there is only partial coverage of reed and other tall emergent vegetation; full coverage may take up to five years.
To the north of the new reed bed is a small, dry field, formerly under agriculturally improved grass but now cultivated with a mixed crop of wild bird seed. The field will be ploughed and re-sown every spring (or perhaps every other spring if biennial crops are used) and it becomes a haven for a wide variety of birds during the autumn and winter months. It attracts not only large numbers of seed eating species such as Goldfinch, Linnet and Chaffinch but also unusually high numbers of insectivorous species such as Dunnocks and various thrushes. The crop has potential for scarce species and has already harboured a few Brambling in its first winter.

Colyford Marsh
Probably the key area of interest for the birdwatcher here is (perhaps ironically) not the nature reserve itself but the land to the east known as Colyford Marsh. This marsh is in private ownership but partly managed by kind permission of the landowner with various measures designed to increase its wildlife value. A large, occasionally tidal lagoon attracts a wide range of species with water levels adjusted where possible to maximise its interest to waders, particularly in the autumn passage period. A quick look at the species accounts in this report will reveal how productive this lagoon can be as it accounts for almost every wader record for Colyford Marsh. A new, much smaller, lagoon was excavated during 2007. While this will take some time to become established, it is hoped that it may become a further area of interest, particularly for passage migrant waders.

Access and Facilities
Visitor facilities at Colyford Common include a small hide and a viewing platform. These both give good views over Colyford Marsh with the hide being the best location to view the brackish lagoons. Note that the hide is relatively small and can become crowded, particularly if there are any scarce species present. The proximity of the adjacent tramway places severe limitations on the size of the hide and this unfortunately makes the possibility of enlarging the hide very remote. Both the hide and the platform are accessed via a boardwalk from the entrance gate. There is presently no access to other parts of the LNR, partly for safety reasons in this tidal area and also to avoid disturbance.
The proposed cycle way from Seaton to Weston-super-Mare will eventually follow the western boundary of Colyford Common, affording views over various additional parts of the marsh. Parking on the roadside at Colyford can be difficult as there is no dedicated parking area for the nature reserve; please park carefully and consider the needs of local residents.

Holyford Woods LNR

The Local Nature Reserve at Holyford Woods covers 24ha (60acres) in a steep-sided valley to the west of Colyford. The woods are owned by East Devon District Council and are leased to the Holyford Woodland Trust (who raised half the money needed to purchase the site). Holyford Woods was the 1000th Local Nature Reserve in England, the declaration being made in 2004.

Holyford Woods (Photo by Paul Glendell)

Habitat and Management
Most of the LNR comprises mature deciduous woodland, the canopy dominated by oak with frequent ash. The understorey is largely of hazel with holly dominating in a few places. A wide range of other native tree and shrub species can also be found. A particular feature of Holyford Woods is the amount of dead wood present throughout the site. With a large number of wind-blown and damaged trees, the opportunities for wildlife here are greatly increased.

At the time the nature reserve was acquired, an area of some 4.5 ha (11 acres) was a mature plantation of Douglas Fir. These conifers had been planted in the 1960s on small fields adjacent to the deciduous woodland. Being of very low wildlife value and having reached economic maturity, the area was clear felled in 2005. This part of the nature reserve has now been left to regenerate naturally, a policy which will ultimately give rise to a new area of native woodland. Parts of the regenerating woodland will be managed in order to leave open glades within the gradually maturing tree cover.

Fauna
Holyford Woods supports a range of typical woodland birds. Notable in a local context are

the high numbers of Marsh Tit breeding here as well as one of the few sites in East Devon with records of all three species of woodpecker.
The site is an exceptional location for bats with eleven species recorded in recent years. Among these are records of Bechstein's Bat, making Holyford Woods one of a tiny handful of known summer sites for Britain's rarest mammal.
Dormice also thrive in parts of the woodland and a number of uncommon invertebrates can be found here, including a thriving population of hornets and good numbers of Silver-washed Fritillary butterfly.

Access
Holyford Woods is approached via a public footpath which leads from one end to the other and connects with a number of permissive paths, giving access to most parts of the site. It is probably best accessed from the A3052 at SY232915.

The Axe Estuary and Seaton Bay Bird Report 2007

Timed Tetrads and Roving Records – Donald Campbell

Last year's report included an article by Gavin Haig on his enthusiasm for sea watching. One of my enthusiasms involves attempting to count common birds, often in 'ordinary' countryside, as part of national surveys organised by the British Trust for Ornithology. These have led to atlases of Breeding Birds (1976), Wintering Birds (1986) and the New Atlas (1993). Another project, using timed tetrad visits and roving recorders started in November 2007. Tetrads – 2km by 2km squares on Ordnance Survey maps – are visited for four hours in winter and for four hours in the breeding season with all birds seen or heard recorded. These are timed tetrad visits, while roving recorders can visit any area adding additional birds. Of some 100,000 tetrads in Britain and Ireland 5 fall entirely within 5km of Axmouth Bridge, 10 are partly within that area and 4 are partly sea.

The Undercliffs NNR tetrads
Being involved with the Undercliffs National Nature Reserve I started in the coastal tetrads east of Axmouth. Before reaching the Undercliffs the flocks of Linnets, Skylarks and Yellowhammers in the stubbles had decreased since October and were to disappear when the fields were ploughed. In the NNR birds were scarce along a rough route through the Chasm, over Goat Island to Rousdon returning by different 'paths' and the Plateau. 15 species including Marsh Tit, Bullfinch and Jay were contacted more than ten times in the two coastal squares SY28U and Z.

The estuary tetrads
The Estuary came next. Tetrads SY29K and L present marked contrasts, for in both an hour by the Axe needs to be paired with an hour amongst the roads and gardens of Seaton, Colyford and Colyton. Counting birds in towns is particularly difficult so I missed the 7 Blackcaps in the back garden of 38 Durley Road!

Four 'ordinary' tetrads
In contrast to counts in the Axe tetrads one might expect those in less varied 'ordinary' areas to be more similar. However for most species in the two inland tetrads east of Axmouth, Bindon and Rousdon and two other standard areas, north of Holyford and around Lower Bruckland the hourly counts were very different. Only two species showed any consistency with Blackbird counts of 26, 15, 23, 33, 28, 13, 13 and 16 while Robins only varied between 9 and 15 per hour. Table 1 shows the ten species with the highest totals in these four tetrads and the count range in frequency per hour.

The Axe Estuary and Seaton Bay Bird Report 2007

Species	Total	Most per hour	Least per hour
Woodpigeon	394	270	3
Chaffinch	213	83	0
Starling	200	150	0
Carrion Crow	197	104	6
Rook	182	74	0
Blackbird	167	33	13
Pheasant	113	37	0
Jackdaw	109	42	0
Skylark	106	82	0
Robin	101	15	9

Table 1. The contrasting counts for ten common species in four 'ordinary' tetrads.

Conspicuous and gregarious species are likeliest to have the highest counts so that when a population estimate is wanted some sort of conversion factor will be needed; this must also take into account the habitats within that tetrad.

Coastal tetrads
The coastal habitat of rocks, pebbles and cliffs between Culverhole and Beer hold few birds. Only 20 Curlew, 14 Rock Pipits, 10 Cormorants, 7 Oystercatchers, 4 Ravens and 3 Peregrines were found along the 6km of coast.

Inland tetrads
Inland, I surveyed all 11 tetrads in SY29 and 28 south of Colyton and mainly, or partly, within the 5km radius. The 86 species found, and roving recorders will no doubt add more, can be loosely divided into 5 groups on the grounds that I maintained at the start that my enthusiasm is for common or widespread species in 'ordinary' countryside. On the evidence of my survey so far Sparrowhawks, like Woodcock and Lesser Redpolls, are decidedly scarce and these birds form one group. The second is made up of birds associated with the estuary or the sea which are not in 'ordinary' country. That leaves three other groups, the noticeable or numerous ones, dominated by corvids (listed in Table 2), five less numerous but equally widespread species and those that are still widespread, occurring in five or more tetrads, but with low populations or being difficult to find.

Table 2. Numbers of the most counted species in 11 inland tetrads.

Species found in all tetrads				Less widespread species	
Woodpigeon	628	Blue Tit	169	Starling	283 in 6 tetrads
Rook	457	Meadow Pipit	165	Redwing	207 in 8 tetrads
Carrion Crow	415	Wren	144	Greenfinch	144 in 9 tetrads
Chaffinch	414	Pied Wagtail	144	Fieldfare	121 in 5 tetrads
Blackbird	393	Pheasant	143	Skylark	114 in 5 tetrads
Jackdaw	342	Great Tit	133	Goldfinch	86 in 9 tetrads
Robin	228	Long-tailed Tit	126		
House Sparrow	227	Goldcrest	108		

Song thrush (90), Dunnock (83), Coal Tit (83), Magpie (46) and Buzzard (35) all occurred in at least ten tetrads whilst ten species remain to form the final group. It will be interesting to find out more about the breeding status of these species which were found in at least five tetrads but with less than 40 individuals as is shown in Table 3.

Species	Number	Tetrads
Bullfinch	32	9
Raven	28	7
Siskin	27	5
Jay	25	7
Marsh Tit	21	6
Nuthatch	29	5
Treecreeper	19	6
Grey Wagtail	18	7
Great Spotted Woodpecker	14	8
Mistle Thrush	12	6

Table 3. Widespread but scarce species in 11 inland tetrads.

There is another series of winter counts to be done but later I will be particularly keen to find evidence of breeding for those species in the fifth group. That is an area where roving recorders can be very helpful and I can provide the appropriate forms or they can be downloaded from www.birdatlas.net . Field work on the Atlas will continue until summer 2011.

Contacts and Links

East Devon District Council
Council Offices, Knowle, Sidmouth, Devon, EX10 8HL, 01395 516551
Countryside Service, address as above, 01395 517557
www.eastdevon.gov.uk/countryside
Nature Reserves Officer, Fraser Rush, 07734 568937
Education Ranger, James Chubb, 07734 568985

Axe Vale and District Conservation Society
Chairman,
Mike Lock, Glen Fern, Whitford Road, Musbury EX13 7AP, 01297 551556
Hon. Secretary, Joan Millard, 15 Grove Hill, Colyton, EX24 6ET, 01297 553447
www.axevaleconservation.co.uk

Devon Bird Watching and Preservation Society
Secretary, Mrs Joy Vaughan, 28 Fern Meadow, Okehampton, Devon, EX20 1PB, 01837 53360
East Devon Branch Secretary, Jonathon Ruscoe, 01404 822689
County Recorder: Mike Tyler, The Acorn, Shute Road, Kilmington, Axminster, EX13 7ST, 01297 34958, devon-birdrecorder@lycos.com
www.devonbirds.org

RSPB South West Regional Office
East Devon Branch Chairman, Jonathon Ruscoe, 01404 822689
Fran Luke, Events/Public Affairs Co-ordinator, Keble House, Southernhay Garden, Exeter, Devon, EX1 1NT, 01392 453758
Crewkerne Group Leader., Mrs Denise Chamings, Daniels Farm, Lower Stratton, South Petherton, TA13 5LP, 01460 240740
www.rspb.org.uk/groups/crewkerne

Devon Wildlife Trust
Cricklepit Mill, Commercial Road, Exeter, Devon, EX2 4AB, 01392 279244
Secretary, East Devon Local Group, Gill Thomas, 01404 861406
www.devonwildlifetrust.org

Axe Estuary Ringing Group
Group Leader, Mike Tyler, The Acorn, Shute Road, Kilmington, Axminster, EX13 7ST, 01297 34958

Axe Estuary Birds
Free twice-monthly email newsletter, contact David Walters on 01297 552616 or email:davidwalters@eclipse.co.uk

Websites of interest

Backwater Birding – Seaton (Forum covering recording area)
http://www.birdforum.net/showthread.php?t=50172

Wildlife in Devon
Webmaster, Paul Boulden, http://www.wildlifeindevon.org.uk/index.html

Axe Valley Birding
Webmaster, Phil Abbott, http://www.axevalleybirding.moonfruit.com/

Portland Bill Bird Observatory
http://www.portlandbirdobs.btinternet.co.uk/

Dawlish Warren National Nature Reserve
http://www.dawlishwarren.co.uk/

Birdguides Ltd
http://www.birdguides.com/

Mike Hughes Illustrator
http://www.mhughesillustration.co.uk/

Spoonbill by Karen Woolley